Nuclear Development

Management of Depleted Uranium

A Joint Report by the
OECD Nuclear Energy Agency
and the
International Atomic Energy Agency

NUCLEAR ENERGY AGENCY
ORGANISATION FOR ECONOMIC CO-OPERATION AND DEVELOPMENT

ORGANISATION FOR ECONOMIC CO-OPERATION AND DEVELOPMENT

Pursuant to Article 1 of the Convention signed in Paris on 14th December 1960, and which came into force on 30th September 1961, the Organisation for Economic Co-operation and Development (OECD) shall promote policies designed:

- to achieve the highest sustainable economic growth and employment and a rising standard of living in Member countries, while maintaining financial stability, and thus to contribute to the development of the world economy;
- to contribute to sound economic expansion in Member as well as non-member countries in the process of economic development; and
- to contribute to the expansion of world trade on a multilateral, non-discriminatory basis in accordance with international obligations.

The original Member countries of the OECD are Austria, Belgium, Canada, Denmark, France, Germany, Greece, Iceland, Ireland, Italy, Luxembourg, the Netherlands, Norway, Portugal, Spain, Sweden, Switzerland, Turkey, the United Kingdom and the United States. The following countries became Members subsequently through accession at the dates indicated hereafter: Japan (28th April 1964), Finland (28th January 1969), Australia (7th June 1971), New Zealand (29th May 1973), Mexico (18th May 1994), the Czech Republic (21st December 1995), Hungary (7th May 1996), Poland (22nd November 1996), Korea (12th December 1996) and the Slovak Republic (14 December 2000). The Commission of the European Communities takes part in the work of the OECD (Article 13 of the OECD Convention).

NUCLEAR ENERGY AGENCY

The OECD Nuclear Energy Agency (NEA) was established on 1st February 1958 under the name of the OEEC European Nuclear Energy Agency. It received its present designation on 20th April 1972, when Japan became its first non-European full Member. NEA membership today consists of 27 OECD Member countries: Australia, Austria, Belgium, Canada, Czech Republic, Denmark, Finland, France, Germany, Greece, Hungary, Iceland, Ireland, Italy, Japan, Luxembourg, Mexico, the Netherlands, Norway, Portugal, Republic of Korea, Spain, Sweden, Switzerland, Turkey, the United Kingdom and the United States. The Commission of the European Communities also takes part in the work of the Agency.

The mission of the NEA is:

- to assist its Member countries in maintaining and further developing, through international co-operation, the scientific, technological and legal bases required for a safe, environmentally friendly and economical use of nuclear energy for peaceful purposes, as well as
- to provide authoritative assessments and to forge common understandings on key issues, as input to government decisions on nuclear energy policy and to broader OECD policy analyses in areas such as energy and sustainable development.

Specific areas of competence of the NEA include safety and regulation of nuclear activities, radioactive waste management, radiological protection, nuclear science, economic and technical analyses of the nuclear fuel cycle, nuclear law and liability, and public information. The NEA Data Bank provides nuclear data and computer program services for participating countries.

In these and related tasks, the NEA works in close collaboration with the International Atomic Energy Agency in Vienna, with which it has a Co-operation Agreement, as well as with other international organisations in the nuclear field.

FOREWORD

In 1999, the OECD Nuclear Energy Agency and the International Atomic Energy Agency jointly decided to establish an expert group to carry out a study on the management of depleted uranium and its potential uses. This expert group aimed to provide a forum to promote a better understanding of the global status, trends and management of depleted uranium, and to exchange information, share experiences and explore potential international collaboration. The results of this study are presented in this report.

This report focuses only on depleted uranium arising as "tails" of uranium enrichment activities. It does not address stocks of depleted uranium that may be obtained from the reprocessing of spent nuclear fuel. Although there are radiation protection, security and safeguard issues related to the management of depleted uranium, they are not specifically addressed in this report. Similarly, economic considerations for potential uses are not elaborated.

The potential use of depleted uranium, as considered within the scope of this report, is focused on peaceful applications; uses of depleted uranium for military purposes are not analysed.

TABLE OF CONTENTS

EXECUTIVE SUMMARY

This report on "Management of Depleted Uranium" presents an overview of the current inventory, potential future arisings, management and options for use of depleted uranium. By definition, depleted uranium contains ^{235}U with a concentration lower than the 0.711% occurring in nature. Since the dawn of the nuclear era, large stocks of depleted uranium have arisen as a by-product of uranium enrichment operations. Currently world stocks of depleted uranium are estimated at around 1.2 million tU. This stockpile has been contained primarily in the form of uranium hexafluoride (UF_6) in metal cylinders and stored at enrichment facilities. Depleted uranium is a potentially valuable resource, with identified uses as an energy source and in shielding and industrial applications. As these uses are currently limited, there is an interest in countries holding such stockpiles in exchanging information on safe storage, potential future uses and disposition options.

The stock of depleted uranium arising from enrichment plant operation is expected to increase by up to 57 000 tU annually for the foreseeable future, though under some planning scenarios, this will be reduced by up to 20% by substitution with down-blended highly enriched uranium and the use of mixed oxide fuel.

Depleted uranium can be safely stored as UF_6 in coated steel containers in outdoor yards provided that there is a container management program to ensure that there is no degradation of the integrity of the containers. The technology for deconversion into U_3O_8 is in commercial operation, and other technologies exist or are under development to convert the UF_6 to other forms. These forms can be safely stored in warehouses. In all cases, the environmental, safety and health impacts on the workers and the general public living around a storage site have to be considered.

To pursue potential uses of depleted uranium, a number of countries are conducting their own R&D programmes, and are interested in international collaboration. Except where these projects involve national sponsorship, such collaboration is most likely to be subject to commercial competition issues or be substantially limited to the exchange of information and experience.

Long-term disposition of depleted uranium would only need to be considered if depleted uranium stocks were determined to have no economic reuse potential.

This report reflects the increasing awareness in countries holding depleted uranium of the need to ensure proper management and to research on the potential uses and disposition options of such materials. Key issues requiring further considerations on depleted uranium were also identified, such as: how much, how long, and in what form should depleted uranium be stored, and implications for final disposition of depleted uranium. Use for re-enrichment should make the need for such final disposition unnecessary for several decades.

I. INTRODUCTION

Most of the nuclear power reactors currently in operation worldwide are fuelled by low enriched uranium produced in enrichment facilities. A by-product of the enrichment operation of natural or reprocessed uranium is depleted uranium or uranium with a ^{235}U content lower than the natural isotopic assay of 0.711%. Typically, depleted uranium generated by the enrichment of natural or reprocessed uranium has a ^{235}U content of less than 0.5%. Large stocks of depleted uranium have arisen as a result of enrichment operations, especially in the United States and the Russian Federation.

Several strategies may be considered for the use and management of depleted uranium. The choice of strategy depends on several factors, including the governmental and business policy, the alternative uses available, the economic value of the material, the regulatory aspects and disposal options, and the international market developments in the front-end of the nuclear fuel cycle. Technological developments in the conversion of uranium compounds as well as in the enrichment processes are also factors impacting on the strategies to be adopted in the future. The assessment of potential uses of depleted uranium is very important for any decision on the evaluation of management options.

In 1999, the OECD Nuclear Energy Agency and the International Atomic Energy Agency decided jointly to organise an expert group for the assessment of the management of depleted uranium and its potential uses. This expert group decided to conduct a study that would allow a better understanding of the global status, trends and management of depleted uranium and that would provide an opportunity for the exchange of information, the sharing of experiences and for potential international collaboration among interested Member countries. The results of this study are presented in this report.

The report outlines current inventories of depleted uranium, trends and potential future arisings, long-term management alternatives, peaceful use options and country programmes. In addition, the report explores ideas for international collaboration and identifies key issues for governments and policy makers to consider.

This report focuses only on depleted uranium arising as "tails" of the uranium enrichment activity. It does not address stocks of depleted uranium resulting from the reprocessing of used nuclear fuel, commonly Magnox fuel (with less than 0.5% ^{235}U) or high burn-up gas-cooled or light water reactor fuel (with a ^{235}U content generally over 0.5%). The quantities of depleted uranium from reprocessing are relatively smaller than the quantities generated by the enrichment operation (less than 5% of total depleted uranium stocks worldwide) and are usually managed in a very different way.

Depleted uranium is slightly radioactive due to its alpha and beta emissions from the uranium isotopes and their decay daughters. Management and potential uses of this material therefore have to be in compliance with radiation protection principles and practices. Depleted uranium also has to be physically protected and may be subjected to international safeguards. Although there are issues in radiation protection, security and safeguards related to the management of depleted uranium, they are not specifically addressed in this report.

The potential use of depleted uranium as considered within the scope of this report is focused on peaceful applications. The use of depleted uranium for military purposes is not in the scope of this report. Similarly, economic considerations for potential uses of depleted uranium are not elaborated in this report.

II. CURRENT INVENTORIES OF DEPLETED URANIUM

Stocks of depleted uranium exist in several locations around the world and in particular in countries which have operated or currently operate uranium enrichment plants. A list of currently operating enrichment facilities and their characteristics (including location, type, capacity and start-up dates) is presented in Annex 2, Table A2. Enrichment plants have been operating since the 1940s and large stocks of depleted uranium (DU) have built up as a result of about half a century of operation of these enrichment plants. Information on the programmes, activities and plans for depleted uranium reported by participant Member countries and organisations is presented in Annex 4.

Depleted uranium is stored in different forms including uranium hexafluoride (UF_6), uranium tetrafluoride (UF_4), uranium oxides (U_3O_8, UO_2 and UO_3), and uranium metal. The characteristics of the different depleted uranium compounds vary from one form to another and may have a significant impact on the management and disposition of this material. The characteristics of depleted uranium and its product forms are summarised in Annex 3.

World stocks of depleted uranium are estimated at around 1.2 million as metric tons of elemental uranium (tU). About 80% of the world inventories are held in the USA and in the Russian Federation, with the vast majority held in the form of depleted uranium hexafluoride (DUF_6). These stocks are listed in Table 1.

United States

The stocks of depleted uranium held by the US Department of Energy (USDOE) arose primarily from the operation of three gaseous diffusion plants (GDPs) built during the 1940s and 1950s. The oldest plant (the K-25 plant at Oak Ridge, Tennessee) was closed in 1985. Operation of the remaining two gas diffusion plants (Paducah, Kentucky and Portsmouth, Ohio) was taken over by the United States Enrichment Corporation (USEC) in 1993. As of June 2000, the United States depleted uranium stocks (including USEC tails) contain about 480 000 tU. The material is stored in the hexafluoride form in 48-inch cylinders stacked horizontally, two high, in open air yards at the sites of the three gas diffusion plants.

Russian Federation

The stocks held by the Russian Federation arose from the operation of four gaseous diffusion plants (all now closed) and their replacement centrifuge enrichment plants from the early 1950s to date. The stocks are estimated at about 460 000 tU[1] mostly (98%) stored as UF_6 in vertical steel containers in open air yards with the remainder as oxides, metal and calcium diuranate.

---

1. Estimate based on the 1996 Uranium Institute Report [1].

Table 1. **Current Stocks of depleted uranium and future plans (as of end of 1999)**

Holder	Current storage form	Stocks in tU	Management Plans
USA			
USDOE and USEC	UF_6 in 48-inch containers on rafts at 3 sites in open air. Containers being repainted & rerafted.	~480 000[a]	Construct 2 deconversion plants to operate by 2005, (technology selection by 2000). Identify uses by 2009.
Russian Federation			
	UF_6 in 2.5 m^3 steel vertical containers, some 48-inch containers in open air yards.	~450 000[b]	Continue storage as UF_6 as future resource. Investigating deconversion technologies.
	Metal and oxides.	~10 000[b]	
France			
COGEMA and Eurodif	U_3O_8 in painted steel 3 m^3 containers, 3 high in warehouses on 2 sites.	~140 000	Continue deconversion and storage as U_3O_8 as future resource. Some quantities recycled as MOX fuel.
	UF_6 in 48-inch containers on rafts in open air.	~50 000	Continue deconversion and storage as U_3O_8 as future resource.
United Kingdom			
BNFL	UF_6 in 48-inch and 0236 containers on rafts in open air at 2 sites. 0236 containers being cleaned, recoated and are being moved indoors.	~30 000	Continue storage as future resource. Investigating deconversion technology to produce metal.
Urenco			
Germany, Netherlands, and UK	UF_6 in 48-inch containers on rafts at 3 sites in open air.	~16 000	Continue storage as UF_6 as future resource. Re-enrich if economic. Storage as U_3O_8 planned in indefinite future.
Japan			
	UF_6 in 30-inch and 48-inch containers, in buildings.	~10 000[c]	Construct deconversion plant to produce uranium oxides to operate in near future (technology selection by 2001). Identify uses and store as future resource.
China			
	UF_6 in 48-inch and 1m^3 steel vertical containers in controlled condition building.	~2 000[d]	Continue storage as UF_6 as future resource.
Republic of Korea			
	UF_6 in 48-inch containers in open air yard.	~200	Continue storage as UF_6 as future resource. Investigate uses.
	Estimated Total	**1 188 200**	

Notes:

a) As of mid-2000.

b) Estimate based on the 1996 Uranium Institute Report [1].

c) As of February 2001.

d) As of end of 2000.

France (COGEMA/Eurodif)

The stocks held by Compagnie générale des matières nucléaires (COGEMA) and Eurodif arise from the operation of the Eurodif gaseous diffusion plant in France since 1979. The stock of 190 000 tU is held as U_3O_8 in about 3 m^3 painted mild steel boxes stocked in warehouses (~75%) and as UF_6 in 48-inch containers in open-air yards. Deconversion of UF_6 stocks to U_3O_8 is undertaken by COGEMA close to the enrichment site.

United Kingdom (BNFL)

The stocks held by British Nuclear Fuel plc (BNFL) mostly arose from the operation of a gaseous diffusion plant (now decommissioned) from the mid 1950s to the early 1980s and the operation of a centrifuge enrichment plant from the mid 1980s which is now operated by Urenco. These stocks, totalling 30 000 tU, are stored in 48-inch containers and 0236-type vertical steel containers in open air yards. There is some additional stock held as metal, oxides, and UF_4.

Japan

The stocks held in Japan arose from the ongoing operation of centrifuge enrichment plants operated by Japan Nuclear Cycle Development Institute (JNC) and Japan Nuclear Fuel Limited (JNFL) since the late 1970s and early 1990s respectively.

The stocks of approximately 10 000 tU are held as UF_6 in 30-inch and 48-inch containers in buildings, as of 2001.

China

The civil stocks of 2 000 tU in China arose from the ongoing operation of the centrifuge enrichment plant operated by China National Nuclear Corporation (CNNC) since the late 1990s. The stocks are held in 48-inch and 1 m^3 steel vertical containers in monitored buildings. Experience of the storage of similar containers under similar conditions elsewhere in China extends back more than 30 years.

Republic of Korea

The Republic of Korea does not have enrichment facilities but depleted uranium stocks are being held by the Korea Atomic Energy Research Institute. The stocks are derived from enrichment activities performed by the US for the Republic of Korea. The current stocks include about 184.8 t as UF_6, 9.6 t as UF_4, and 3.8 t as metal.

Urenco (sites in Germany, Netherlands, UK)

The stocks held by Urenco arose from the ongoing operation of centrifuge enrichment plants in the UK and the Netherlands from the mid 1970s and in Germany from the mid 1980s.

The stocks of 16 000 tU are held almost entirely as UF_6 in 48-inch containers in open air yards at the three sites.

Other stocks

Depleted uranium has also arisen under various other programmes around the world. For example, South Africa operated two enrichment plants, closed by March 1995, based on the jet nozzle and helikon processes respectively which resulted in 2 200 t of depleted uranium. Most of the South African DU stock was sold on the international market. As of the end of 1999, the total South African inventory is estimated at 73 t in various forms mostly metals and oxides (powders). There are about 4 t stored as UF_6. Oxides are stored in containers and the metal open stacked as ingots inside facilities. Some of the DU metal is used for the fabrication of containers for shipping of commercial radioisotope products, mostly for exports.[2]

Other enrichment activities have resulted in tails arisings not included above for which there are no reliable figures available.

III. TRENDS

Future arisings of depleted uranium

There is currently around 50% over-capacity in enrichment services compared with the primary market demand. New centrifuge based capacity, however, continues to be commissioned. This over-capacity is also currently exacerbated by the substitution of primary enrichment by down-blending of ex-weapons Highly Enriched Uranium (HEU). The main enrichment, and thus depleted uranium or tails generation, in the foreseeable future will come from the enrichment plants listed in Annex 2, Table A2, which includes plants in China, France, Germany, Japan, the Netherlands, the Russian Federation, the UK and the USA.

The primary factor governing future arisings of depleted uranium is the enrichment needs of the installed nuclear generating capacity worldwide. Over the next 15 years such capacity is forecast to show a modest increase as commissioning of new plants exceeds decommissioning. Over the longer term the construction of new and replacement capacity is dependent upon the relative economics of nuclear generation versus alternative sources, the public perception of the threat of global warming and public acceptance of nuclear generation.

An estimate of an ongoing primary demand for uranium of 65 000 tU/a and enrichment requirements of 36.5 million SWU/a (assuming 0.3% ^{235}U tails assay), equivalent to an installed generating capacity of the order of 375 GWe, is taken as a reference case for this study. For this scenario the resultant primary tails are of the order of 57 000 tU/a arising in those countries undertaking the enrichment.

Beyond the primary demand arising from nuclear power plants, there are two other main factors that will affect the volume of depleted uranium generated; they are:

- the use of secondary supplies to displace primary enrichment, these include the use of plutonium from reprocessed fuel and excess military supplies in Mixed Oxide fuel (MOX), and down-blended excess military HEU.

- the operational tails assay of the enrichment plants, which will be affected by factors including the relative price of U_3O_8 and enrichment services.

Secondary supplies

HEU

The down-blending of excess military HEU to produce Low Enriched Uranium (LEU) for use in commercial reactors is a current reality in both the Russian Federation and the USA. Current planning calls for the down-blending of 30 t/a of Russian HEU until 2013 and 10 t/a of (lower mean assay)

American HEU well into this decade. Dependent upon future political developments the duration of these campaigns could be lengthened or shortened. The current effect is to reduce the primary enrichment reactor requirements and thus new depleted uranium (tails) production by around 15%.

The down-blending of the HEU to produce commercial reactor grade LEU requires the production of 1.5% material from tails (to meet the minor isotope specifications). To down-blend the planned 30 tU/a of Russian HEU requires on the order of 850 tU at 1.5% to produce 4.6% product. The tails consumed for this are roughly 1.5% of the primary production. This will also reduce the assay of between 5 950 tU and 11 000 tU of the remaining secondary tails from 0.3-0.2% to 0.1% ^{235}U.

MOX

As a result of the operation of reprocessing plants and related MOX fabrication facilities up to 5% of the world demand for reactor primary enrichment requirements could be displaced by the use of MOX. This would have a concomitant 5% reduction in depleted uranium production.

The potential use of ex-military plutonium in MOX is currently under consideration by the Russian Federation and the USA. A bilateral agreement designed with this objective was signed by these two nations in September 2000. About 3 t/a of plutonium incorporated into MOX would displace 1% of the primary enrichment requirements.

Reprocessed uranium and stockpiles

The use of reprocessed uranium, as with the use of MOX, is dependent upon the operation of reprocessing plants in France, Japan and the UK. If these all operate fully and the uranium is re-enriched, this would result in a reduction of less than 1% in the primary enrichment requirements and tails arisings (assuming 0.8% reprocessed uranium enriched to 3.65%, as opposed to natural enriched to 3.5%, both at 0.3% tails).

Alternatively, the material could be blended down with HEU to produce reactor grade LEU. In this case, the reduction in primary enrichment and tails arisings would be case dependent.

Current stockpiles of enriched product are assumed to be maintained. Any changes in tails production resulting from likely changes in average levels will be insignificant compared with the total production over the next decades.

Enrichment plant operational tails assay

The economics of the nuclear generator define the nominal product and tails assays and volume of product necessary for each reactor. A prime factor in this calculation is the relative cost of enrichment services and of U_3O_8. The higher the cost of U_3O_8 the lower the optimum tails assay at a given enrichment cost. To a first order this dictates the volume of tails arising. However, the operator of the enrichment plant may, for economic, political or capacity planning reasons, choose to operate at a higher or lower mean tails assays temporarily or long term.

The tails from primary commercial operation of enrichment plants has typically ranged between 0.25-0.35% ^{235}U. Compared with the current assumption of a tails assay of 0.3% ^{235}U this range of assays varies the volume of depleted uranium produced by -10% to +14%. While the price of U_3O_8 is depressed by the current oversupply from secondary supplies, the optimum tails assay will remain above 0.3% ^{235}U. In the longer term, as the reserves of low cost U_3O_8 are exhausted and the price rises, the optimum tails assay should return below the 0.3% level. The operator's response, however, is not definable and their operating assay will remain a commercially confidential matter.

Summary

In the absence of secondary supplies, it can be assumed that the generation of primary depleted uranium would increase modestly or would remain at about the current nominal level of 57 000 tU/a. This level, however, could be reduced by up to 20% due to displacement by MOX fuel and down-blended HEU reducing annual primary tails arisings down to the order of 45 000 tU/a whilst these processes continue.

IV. LONG-TERM MANAGEMENT

Strategy

The strategy for the long-term management of depleted uranium is based on the consideration that depleted uranium is a valuable material, which may have various applications, and is not considered a waste. This strategy depends on a number of factors, but primarily the future development of nuclear power programmes around the world. At present, light water reactors dominate nuclear power generation and will continue to do so for the coming decades as development programmes for fast reactors have been delayed or even abandoned in many countries. It is, however, likely that very long-term use of nuclear energy would ultimately require the implementation of fast reactors. In this case, depleted uranium would become an important energy source.

Similarly, the depleted uranium recoverable by re-enrichment is a potentially valuable source of ^{235}U for light water reactors, whilst the remaining ^{238}U could be used for the "blankets" of the fast reactors. In the absence of these, or other, large-scale applications, however, final disposition in some form of repository would have to be considered. In the meantime, the principal factor determining the strategy is the economic storage of this material in safe and ecologically friendly facilities that would ensure its availability for future use.

Timescale

Under current conditions, the time scale for the management of depleted uranium is defined by the economics of re-enrichment in relation to mining primary uranium resources. For the latter, the 1999 IAEA/NEA *Uranium: Resources, Production and Demand* [3] report identifies "Reasonably Assured Resources" and "Estimated Additional Resources" of at least 1 254 000 tU as of 1 January 1999 recoverable at ≤ USD40/kgU. With an annual reactor related uranium requirement of 65 000 tU/a these primary resources would last at least 20 years. Due to the availability of "civilian" material in stockpiles, the disposition of surplus defence material and the use of reprocessed power reactor fuel uranium and plutonium, these primary resources will last considerably longer. Current data also indicate a total of 3 000 000 tU available recoverable at a cost up to USD80/kgU, sufficient for a total of 46 years of reactor operations. However, it is likely that at some point within the next 50 years, the first re-enrichment of tails will become economic compared with the mining of primary resources.[2] The secondary tails will still contain significant quantities of recoverable ^{235}U and with only limited further resources of U recoverable at below USD130/kgU a second re-enrichment could well become economic within a further two to three decades.

2. 1kg U at 0.711% ^{235}U requires 3.894 kgU feed at 0.3% ^{235}U and 1 SWU leaving secondary tails at 0.158% ^{235}U.

The depleted uranium management requirements are thus dominated by the need for economic storage for the next seven to ten decades to ensure its availability for re-enrichment to replace and supplement primary production. This period also covers the time necessary for the development of fast reactors in the case deployment of this technology becomes evident.

Storage as UF_6

Technically, depleted UF_6 can be stored safely for many decades in painted steel containers in the open air in storage yards. Internal corrosion does not represent a problem. Limiting factors are the external corrosion of the containers and the integrity of the "connection" seals. However, the influence of these factors can be minimised with an adequate preventive maintenance programme.

External corrosion

There are three primary causes of external corrosion, all of which are preventable:

- Standing water contact

  Contact with ground water can lead to localised corrosion and ultimately perforation of the container wall. This can be prevented by correct design and construction of the storage yard to ensure good drainage and/or by ensuring that the structural part of the container is raised above "ground" level. Periodic inspection is necessary to ensure the stability of the container support structures and drainage of the yard.

- Handling damage

  Damage to the container during handling, for example, during plant operation, during transfer to, or whilst stored in, the storage yard, can result in damage to the (protective) paint coating and to the container structure itself. Both can result in initiation sites for surface corrosion. Prompt treatment and repainting of the area will prevent surface rust from developing.

- General ageing of paint

  Coated surfaces in the open air are subject to the whole range of atmospheric conditions, wind erosion and natural phenomena. These conditions lead to a slow degradation of the integrity of the paint layer eventually leading to loss of coating in local areas. These areas form the sites for the start of localised corrosion that would spread with time. Periodic inspection coupled with prompt treatment and repainting of these areas will prevent surface rust from developing.

Storage of containers in buildings in which the humidity is monitored and condensation on containers is prevented reduces further the inspection and regular localised treatment and repainting work.

Historic holdings of UF_6 in containers which have not been regularly inspected and in which any localised surface rust has not been treated and repainted form a special group. In principle, provided that the container wall is not perforated with rust or there are no localised areas where corrosion has significantly reduced the wall thickness, these containers can be cleaned/treated and recoated and replaced in the storage yards, then can have an indefinite storage life. However, where a container has suffered damage or is otherwise beyond treatment, the contents will need to be transferred to a new container.

Connections seals

All containers have some form of connection, and the long-term integrity of the seal of such connections is equally as important as the absence of corrosion for the safe storage of UF_6. Failure of this seal would lead generally to very localised contamination as the UF_6 reacts with localised moisture forming solid reaction products and hydrogen fluoride (HF). Periodic inspection would detect these localised solid reaction products. Experience to date indicates that failure of plugs/seals etc. has not been a limiting factor in storage anywhere.

Experience

There is a widespread experience of the storage of UF_6 in steel containers in open-air storage yards. Even without routine treatment of localised corrosion, containers have maintained structural integrity for more than 50 years. The most extreme conditions experienced have been in the Russian Federation (Siberia) where temperatures ranged from +40°C to -40°C and from deep snow to full sun. The most protective storage has been in China where containers have been stored in controlled conditions indoors for more than 30 years. Retreatment of containers stored outdoors for long periods is being undertaken both in the USA and UK. Storage experience in the United States has proved to be satisfactory with only a small number of minor leaks.

Deconversion

Depleted uranium stored in the UF_6 form represents a potential chemical hazard if not properly managed. Alternatives for the strategic management of depleted uranium therefore include the deconversion of UF_6 stocks to stable forms more suitable for long-term management. Due to their high chemical stability and low solubility, uranium oxides in general are favoured forms for this. High density UO_2 and U metal are other possible forms with a high long-term stability. [4]

An industrial scale facility is necessary for any significant deconversion programme. The necessary technology and processes have already been developed. Currently there is only one commercial organisation (COGEMA in France) with an ongoing programme to deconvert UF_6 to U_3O_8. This programme uses kiln technology developed in parallel with the diffusion plant with the aqueous HF by-product being sold. Other organisations have investigated similar technologies or are developing alternative technologies to deconvert UF_6 to a stable oxide, UF_4, or metal form. However, whatever technology is deployed, deconversion will significantly reduce a potential chemical hazard but will adversely affect the economics of re-enrichment and delay the time at which the material can be used.

Deconversion technologies are commercially sensitive. The location of any deconversion facility will have to take into consideration the condition of any containers used for storage and the transport safety issues.

Storage as U_3O_8

Generally, though not exclusively, storage as U_3O_8, the most stable oxide, is considered for long-term storage where continued storage as UF_6 is not appropriate. This has the advantage of reducing the toxicological and environmental risks in case of accidents by external impact, plane crashes and fires. This can be undertaken in painted mild steel containers stored in a "warehouse". Provided condensation and structural damage is prevented, there is no effective technical limit to the duration of the storage in this form. Routine inspection and structural maintenance of the building structure will be necessary.

Experience

Large-scale storage of compacted U_3O_8 powder has been undertaken in France since 1984. About 140 000 tU of UF_6 have been converted into U_3O_8. The powder is held in about 3 m^3 painted mild steel DV 70 type containers stacked three high in "warehouses." The containers are now also being transported by rail to a new storage site (Bessines) in Central France.

Storage of other compounds

Large volumes of UO_3 and UO_2 and small volumes of UF_4 have been stored in barrels in storage facilities. There is no effective technical limit to the duration of such storage, provided that fluoride carryover during the deconversion process is prevented so that HF does not subsequently form, and that condensation is prevented and structural damage avoided during handling and storage. In addition bulk uranium metal has been stored for long periods both in containers and as itself. This represents the most compact form of storage.

Other storage considerations

Three factors significantly affect the storage of depleted uranium:

- Chemical safety

  The toxicological and environmental risk from the storage of depleted uranium is greatest when it is stored as UF_6 and is associated with its interaction with moisture to produce HF. The capacity of UF_6 storage yards can thus be limited by the "safety case" represented by the worst case "catastrophe" relevant to the site and the location of the site with respect to the local population. The limiting factor is normally fire and the risk of aircraft crashes, and the yards should thus be kept free from inflammable material.

- Radiation safety

  There is no criticality safety consideration for the storage of depleted uranium. The mass of material normally involved, however, does lead to the need for consideration of the radiation dose received, both by site staff and off site.

  This can influence the design and the location of the storage yard within a site and with respect to the site boundary and any "local" use of the land surrounding the site. These factors can also affect the design of any building housing UF_6 and U_3O_8, metal or other compounds containing depleted uranium.

- Security

  Although management of depleted uranium includes the need for physical security, depleted uranium does not represent a significant proliferation risk, nor is it an attractive ecological terrorist target.

Final disposition

It is not envisaged that final disposition of the depleted uranium will be necessary within the next 70-100 years while nuclear generation of electricity continues. Should fast reactors become a significant factor in future energy scenarios, this time factor will be significantly extended or the need reduced. However, decisions to phase out nuclear power generation could shorten this time-scale. The design of final repositories, in particular the specific geological structure in each case, will ultimately define the requirements of the packaging and acceptable chemical forms of the depleted uranium.

Current management plans

Prime enrichers including Eurodif, the Russian Federation, Urenco, Japan and China, and two static stockpile holders, BNFL and the Republic of Korea, consider the depleted uranium as a valuable future energy resource requiring economic long-term storage as UF_6 or primarily as U_3O_8 when site constraints or politics so dictate. Current plans for future management in reporting countries and organisations are listed in Table 1. Additional information on programmes and plans is presented in Annex 4.

- COGEMA/Eurodif: The current policy of deconversion to U_3O_8 at a rate slightly higher than current tails production, thus reducing the stock held as UF_6 in current facilities, will be continued. The U_3O_8 will be stored for the indefinite future until required.

- BNFL: The current maintenance programme on 0236 type containers will be completed and these are being moved to storage indoors. A plasma based deconversion process is being investigated to produce uranium metal for storage and to utilise the resultant elemental fluorine for industrial applications. Potential applications for the depleted uranium will be investigated if, as and when they are identified.

- Urenco: The current programme of storage as UF_6 with inspection and preventive maintenance will be continued. The stock will continue to be re-enriched where economically achievable. Licences for a U_3O_8 store have been received in the Netherlands (COVRA) and an application is in preparation in Germany as a contingency against future needs. Initial design work for above ground U_3O_8 stores has been started.

- Japan: The basic plan is to pursue beneficial use of DU, but, on the other hand, deconversion to uranium oxides is under consideration at JNC for long-term storage.

- Russian Federation: The current programme of storage as UF_6 with inspection and preventive maintenance at each of the four sites will be continued. The strategy for the management of the depleted uranium stocks is currently under review to look at deconversion technologies, the deconverted chemical form (UF_4, oxides or metal), their storage, management and use, and, finally, the start of bulk processing by 2010. Finance is likely to be a problem for the execution of this programme.

- China: The current programme of storage as UF_6 will be continued.

- Republic of Korea: The current storage as UF_6 will be continued; the deconversion plant will remain mothballed. Small-scale applications will be pursued where economically feasible.

- USDOE: The current container repainting and re-rafting programme will be completed. One or more deconversion technologies will be selected and deconversion plants sited at the currently operating USEC gas diffusion plants (enrichment sites). These should be operating by 2005. The USDOE is pursuing a programme to identify beneficial uses for its stockpile by 2009. In contrast with the majority of enrichers, the United States does not view its stocks as a future energy resource and is pursuing a policy of deconversion to a more stable form with a goal to disposition while it explores potential beneficial uses for the material.

Summary

Depleted uranium arising from the operations of enrichment plants can be safely stored as UF_6 in coated steel containers in external yards, provided that contact with standing water is prevented and that containers are routinely inspected and localised defects leading to corrosion are treated.

The capacity of storage yards is determined by site specific constraints based on the "risk assessment" of a catastrophic event resulting from the chemical toxicity of UF_6 and the radiation dose received by staff and at site boundaries.

Should deconversion to a more chemically inert form be necessary or convenient, commercially proven technology is available to produce U_3O_8. Other technologies are available or under development that could produce UF_4, oxides or metal.

With suitable maintenance, these inert compounds can be stored in warehouses to optimise the economics of re-enrichment and permit other uses over the coming seven to ten decades.

V. OPTIONS FOR USE

A number of potential applications have been identified for the use of depleted uranium worldwide. In general, these uses can be classified into nuclear and non-nuclear uses.

Nuclear uses

The depleted uranium stocks represent a major reserve of energy usable by re-enrichment to displace primary mined ore, by blending with HEU to form reactor grade LEU, blending with plutonium to form MOX, as a blanket material in a fast reactor, or as a minor constituent in some PWR and CANDU fuel designs.

Depleted uranium is an excellent shield against nuclear radiation and thus has potential applications in the nuclear, medical and other industries where biological shielding is necessary. This can be achieved by the incorporation of depleted uranium metal liners, e.g. Cermet encased in steel, the use of depleted uranium oxide as an aggregate in concrete (Ducrete), or as a component of materials inside a waste package.

Energy resource

Re-enrichment

The timing for the re-enrichment of depleted uranium is determined by the economics. Currently re-enrichment is only economic in centrifuge enrichment plants with spare capacity where operating costs are low. The current depleted uranium stockpile of 1.2 million tonnes U (0.3% ^{235}U) would provide 336 000 tU of equivalent natural uranium, leaving 864 000 tU of secondary tails with an assay of 0.14%, sufficient for 5 years of operation of the world's nuclear reactors at current uranium requirements. The secondary tails could then be re-enriched providing a further 106 000 tU equivalent to another one and a half years of world's reactor requirements leaving 758 000 tU of tertiary tails with an assay of 0.06% ^{235}U.

The Russian Federation has used and continues to utilise spare enrichment capacity in its enrichment plants for re-enrichment of tails. However, it will only become economic to construct plants specifically for re-enrichment of tails or depleted uranium when the price of 1 kg U_3O_8 plus its conversion costs to UF_6 are higher than the cost of one SWU.

Down-blending HEU

The down-blending of HEU from decommissioning of weapons requires the use of (1.5%) LEU produced from tails to meet the ASTM specification for the minor isotopes. The dilution factor to produce reactor LEU at 4.4% ^{235}U, obtained by blending 1.5% enriched LEU with HEU is 30:1 for 90% ^{235}U, HEU.

Using the 1.5% LEU produced from the tails stocks will thus reduce the stockpile by 30 kgU for every kilogram of HEU down-blended. The dilution factor reduces as the assay of the HEU reduces, e.g. to approximately 16:1 with HEU at 50% ^{235}U.

The total potential use of depleted uranium for this use is thus dependent upon the political decision regarding the decommissioning of weapons based on HEU and the actual assay of the material.

The Russian Federation and the USA have an intergovernmental agreement for the period ending in 2013 which governs the down-blending of 30 t HEU annually in the Russian Federation. This programme will consume 900 tU/a from tails stocks.

Mixed oxide (MOX) fuel

MOX fuel is a mixture of plutonium dioxide and uranium dioxide. The uranium dioxide is usually depleted uranium. Typically, MOX fuel comprises 6% Pu and 94% DU. MOX fuel assemblies can be currently loaded in light water reactors with uranium fuel assemblies in the ratio of 1:2. Hence for a 900 MWe reactor, 16 out of 52 assemblies per reload will be MOX. The 16 fuel assemblies require 390 kg Pu and 7 t DU. The Uranium Institute forecasts demand for depleted uranium in MOX fuel to increase to around 340 tU/a by 2003 and then remain stable at this level.[1]

Fast reactor fuel

In a fast reactor, the core consists of UO_2–PuO_2 with a blanket around the core of either natural or depleted UO_2.

For fast reactors, the Uranium Institute has estimated global average annual demand to be around 100 t of DU in 1996, if they are operated as breeders (this figure included Superphenix, now closed down). If fast reactors are operated to target a neutral plutonium balance, demand of DU is reduced by around 33%, and if these reactors are operated to consume plutonium, by 66%.

Consumption of DU as fast reactor fuel is expected to be small unless fast reactors become commercially viable. This is unlikely before 2030.

Light-water reactor fuel and CANDU [1]

In some designs of PWR fuel, DU is used in fuel rods containing burnable absorber (an example of which is gadolinium oxide). Only a few of all fuel rods in each fuel bundle are of this kind and only about half of all fuel bundles contain such rods.

The total world use for DU in some kind of light water reactor fuel design is estimated at 5 t/a.

In the initial core for a CANDU reactor there are a number of bundles of DU. If there is a fuel failure, this fuel may be replaced with DU fuel. The estimated use of DU for CANDU reactors is 5 t/a. This figure is likely to remain unchanged over the period to 2010.

Surplus plutonium

In September 2000, the United States and the Russian Federation signed a bilateral agreement on surplus plutonium disposition. The agreement calls for 34 t of Russian military grade plutonium to be used in reactors with the first production of MOX fuel in 2007. It is anticipated that an amount of depleted uranium will be used in the manufacture of this fuel.

Under this agreement, the United States intends to dispose 25.5 t of surplus weapons grade plutonium by making MOX fuel that will be burned in commercial reactors to produce electricity. The United States is planning to use a total of about 700 t of depleted uranium beginning in 2007 and continuing through 2019, or about 60 t/a to manufacture this fuel. The United States also intends to dispose 8.5 t of some surplus plutonium by immobilising it within a ceramic matrix and then disposing of it in a repository. Additionally, approximately 4 t of less-than weapons grade plutonium will be dispositioned outside the agreement using immobilisation. This form of plutonium disposal would use a maximum of about 26 t of depleted uranium beginning in 2007 and continuing through 2019, or about 2.2 t/a.

Shielding applications

Engineered shielding materials

A variety of materials exist or have been postulated that combine depleted uranium aggregate, typically in the form of dioxide, with a matrix or binding agent to produce a high-density shielding material. Matrices that have been proposed include cement, polyethelene, and graphite-based materials. These materials would have much greater shielding capability per unit thickness and unit shield weight than less costly materials such as standard concrete. This attribute could be cost-effective in some applications.

Biological shielding

DU provides shielding against X-rays and gamma rays for radiation protection. DU products are regularly used in making products such as gamma defectoscopes, gamma radiographic units for material testing, shielding in radiotherapy appliances, diaphragm-system shielding walls, shipping containers, research and medical equipment that requires reliable biological shielding of personnel.

The Kapline report [5] indicates that there could be significant potential for DU consumption in shielded rooms either by enhancement of lead and cement in existing rooms or the use of DU in new rooms and in mobile shielded rooms. Mobile rooms are rented to hospitals or used during the construction of permanent facilities.

Repository applications [5]

The potential uses for DU in nuclear waste repositories are containers, waste package fill and down blending of HEU reactor fuel.

Containers – in some designs the containers will have to shield personnel from the waste and also help isolate the waste from the environment. The DU could be used as a shielding material in the containers. Several forms of DU shielding could fit this function, e.g. Ducrete, DU Cermets.

It has been proposed [6] that spent nuclear fuel waste package structural components and the internal basket structures be constructed of a depleted uranium dioxide (DUO_2) steel cermet. The cermet consists of DUO_2 particulates embedded in steel. The exterior of the waste package would have a layer of corrosion-resistant material chosen to maximise corrosion resistance in the particular geological environment. The DUO_2 steel cermet may (a) reduce the long-term potential for nuclear criticality in a repository containing spent fuel and (b) reduce radionuclide releases by reducing the driving force for dissolution of spent fuel while beneficially using excess depleted uranium.

Waste packaging fill – fill materials, such as DU dioxide particulates or DU borosilicate glass beads, could be used in waste packages. These fillers would eliminate void space to prevent eventual crushing of the waste package, shield workers from the radiation, provide a barrier against release of radionuclides from the waste package and improve heat transfer of radioactive decay heat from the waste to reduce thermal degradation of the waste form. There are still technical concerns about the suitability of DU for this application.

In order to maintain subcriticality in a repository depleted uranium may be used as a blending material to reduce the enrichment of irradiated research reactor fuel which is not reprocessed for technical or economic reasons.

Non nuclear uses

Depleted uranium is a low cost material that is readily available. It has high density, high melting point, high tensile strength and it lends itself to simple mechanical treatment. DU can be used to produce articles with complex configurations by means of casting and stamping and in combination with stainless steel it has a high resistance to corrosion.

Counterweights

Because of its density, depleted uranium provides a significant mass in a small area where volume constraints prevent the use of less dense materials. Concerns about using the material in public areas due to the low radiation and chemotoxicity is however a potential limitation and substitution by other high density, less toxic metals, e.g. tungsten is often considered.

DU counterweights are used in the transport industry to compensate for shifting cargo loads or fuel consumption and are being investigated as a replacement for lead in elevator, crane and fork-lift truck applications. DU counterbalances are also used in control surfaces (elevators, ailerons) in wide body aircraft, for vibration absorption, centrifugal counterweights and in gyrocompasses.

Lamp filaments

A small tube of uranium dioxide has been used to eliminate sudden surges of current through bulbs when the lights are turned on, thus extending their lives.

Oil industry

DU can be melted and moulded into appropriate shapes and has high density and tensile strength. DU is being investigated for use in drill collars, well penetrators and well shape perforators for use in the oil well drilling industry. The key issue for the oil industry use is performance and, given the natural radioactivity found in oil wells, DU use does not represent a major concern.

Energy storage flywheels

The Kapline report [5] indicates that there are two markets for flywheels – portable (electric vehicles) and stationary (power plants). DU is more likely to be used in power plants than in electric vehicles. The high density of DU metal makes it a potential candidate material for flywheels used for inertial energy-storage devices where space is limited.

Catalysts

There has been some research on the application of DU as a catalyst for destruction of volatile organic chemicals in off-gas streams [7] and such catalysts have apparently been used industrially for many years.[8] Research on use of DU oxides as catalysts is presently ongoing in the USA.[9,10]

If depleted uranium catalysts were to be in contact with materials that eventually became components of commercial or consumer products there is the possibility of some contamination by the DU and the existence or extent of this carryover would need to be defined. The potential exists that continued research into catalyst supports may obviate this concern.

Semi-conductors

Uranium oxides have characteristics that could potentially result in better performance of currently available semiconductor materials such as silicon, which are reaching their theoretical limits. Application of uranium-based semiconductors might include solar cells and thermoelectric generators. There has been limited study of such use based primarily on existing literature but also including recent experiments, both of which show promise [9, 11]. However, considerable research will be needed to develop near-optimal uranium oxide semiconductors and evaluate their performance and cost-effectiveness.

Other potential future uses

DU could also have potential uses as phosphors, batteries, hydrogen storage, and as an alloying constituent similar to thorium, which has been used for many years. However, these uses have not been studied and it is unclear if applications are practical or would provide significant advantage.

In addition, DU could be used for the production of fluorine. Hydrofluoric acid, HF gas, and/or elemental fluorine are by-products of depleted UF_6 conversion into oxides, metal, etc. HF has uses for CFCs, in the oil, chemical, and nuclear industries among others.

Potential constraints on depleted uranium usage

Some of the constraints on use of DU are the following:

- Any use of depleted uranium has to be in compliance with the basic principles of radiation protection.
- Requirements for tracking and limitations on disposal of products containing DU may be imposed by individual countries.

- Any company, not currently handling radioactive materials, using depleted uranium for the first time will have to learn to meet the obligations of a wide range of national and international regulations covering the production, handling, use, storage, transport and disposal of radioactive materials.
- Manufacturers need to deal with the hazards that result from DU's pyrophoricity, chemotoxicity and radiotoxicity.
- The use of DU will have to demonstrate major cost savings or technical advantages to manufacturers and end users to encourage displacement of existing products or development of new products.
- There is a lack of public acceptance for the use of DU in consumer products.

Prospects for depleted uranium use

At the present time re-enrichment to displace natural uranium and for blending with HEU is the most significant use for DU. Although this application creates further depleted uranium at lower ^{235}U, the net result is a reduction in the stockpile and a saving in natural resources. The magnitude and timing of this use is determined by economic factors. Volumes will only increase significantly when the costs of freshly mined U_3O_8 increase making re-enrichment more economic.

Other uses in the nuclear field including fast reactors, LWR and CANDU fuel and MOX fuel will continue to require modest volumes of DU from the stockpile whilst these reactors remain in operation. Significant increases in use would arise should the fast reactor becomes a widely adopted technology in the coming decades. In the long-term, there is also the potential for the need for significant quantities in repository, shielded cask and shielded room applications. Outside the nuclear energy uses and shielding applications, usage is limited by public acceptability, practical problems associated with licensing and regulatory factors and manufacturing factors associated with its pyrophoricity, radiotoxicity and chemotoxicity.

In all cases, costs associated with the storage of DU should be minimised to optimise the chances for future economic applications.

Summary of research and development (R&D) activities

To pursue potential uses for depleted uranium, a number of countries and commercial organisations are conducting research and development. These countries include the USA, Japan, the Russian Federation, Republic of Korea and China. Countries with no specific research programmes include France, UK and Germany. The following is a summary of the information reported by participant countries and organisations.

National programmes

United States

The US Department of Energy initiated the relatively small "Depleted Uranium Uses Research and Development Programme" in 1999 to explore the potential beneficial uses of the DU, fluorine, and empty carbon steel UF_6 storage cylinders to achieve cost savings to the government. The United States will also carry out research activities necessary to assure the direct disposal of these materials if cost-effective and realistic beneficial uses are not found. The objective of the programme is that by

2009, innovative uses for the products will have been developed that utilise a significant amount of the depleted uranium inventory such that revenue is generated that allows the cost of managing the remaining inventory to be significantly reduced.

Areas of interest to the United States that are being considered for research and development are potential uses of depleted uranium as a shielding material, a component of a high-level waste repository, catalyst, semi-conductor material, and as a phosphor.

Japan

The Japan Nuclear Cycle Development Institute has initiated a feasibility study on usage of depleted uranium. In addition to evaluating depleted uranium's use as fuel for the fast breeder reactor, the Institute is actively investigating uses for depleted uranium as a:

- Shielding material for spent nuclear fuels storage and shipping casks.
- Hydrogen storage alloy that could be used for energy storage when used in conjunction with a fuel cell.
- Magnetic material as a substitute for rare-earth magnetic materials.
- Reduction-oxidation flow cell material that would store electrical energy.

Russian Federation

The Russian Federation has initiated a programme to conduct scientific research and development related to the continued long-term storage of DUF_6 and is investigating the conversion of the UF_6 to other forms using a plasma process. A focus of the Russian research is to create a fluorine cycle within the nuclear industry so as to eliminate the need to introduce virgin fluorine, as well as producing a variety of fluorine compounds for a variety of industrial uses, e.g. freon substitutes (khladone) or hydrogen fluoride.

Republic of Korea

Since the 1980s, the Republic of Korea has been conducting research activities directed to define alternative uses for depleted uranium. A facility to convert UF_6 to UF_4 was constructed during 1987-1991 and operated for some years. A facility for the reduction of UF_4 to DU metal operated since 1984 but was decommissioned in 1994. Another DU metal working facility has been operating since 1983. Research objectives include the development of radiation shielding containers (for a variety of specialised uses e.g. transportation containers for medical isotopes), development of nuclear fuel fabrication technology using DU and fundamental research of liquid metal cooled nuclear reactor fuel.

The Republic of Korea has been interested in investigating the optimal storage form and configuration for its depleted uranium stocks. The objective is to allow the relocation of these stocks currently stored in the United States in preparation for its potential future use in a breeder reactor programme. The Korea Atomic Energy Research Institute (KAERI) is interested in alternative uses of depleted uranium and is seeking an adequate long-term storage technology.

China

China has a programme to investigate the use of depleted uranium in fast breeder reactors but is not currently investigating other uses of depleted uranium.

Commercial programmes

BNFL

A plasma-based deconversion process is being investigated to recover the fluorine component from depleted uranium as high value elemental fluorine gas and high density metallic uranium to minimise storage volumes.

COGEMA

COGEMA has designed and built a facility to use DU in the form of UO_2 for the manufacturing of MOX fuel, which is currently used in European reactors. No other uses for depleted uranium are presently investigated.

Others

Particularly in the United States, many independent small businesses and individuals are conducting limited research to identify potential uses for depleted uranium. For example, a review of the United States Patent Office records reveals that 10 patents were issued during the period 1999-2000 for inventions that represent uses of depleted uranium. The extent to which this is occurring in other countries is not known.

VI. INTERNATIONAL COLLABORATION

Management of depleted uranium is the responsibility of each entity (country and/or company) holding its stock. The strategies for management of the stocks and current arisings of DU are influenced by national and company policies and prospects on nuclear power. Similarly, potential applications outside the nuclear area depend on national views on nuclear power and radiation protection and public acceptance in these countries. Some issues affecting depleted uranium such as, storage and long-term disposition are of common interest and mutual concerns to the countries and companies. There may be areas where international collaboration could be of benefit to all parties involved.

Exchange on information on best practices for the storage of DU

Regardless of the country's strategy for management of depleted uranium, safe and secure storage of depleted uranium is a pre-requisite. Here, an on-going international collaborative effort in sharing information and experience on management of the materials to minimise the environmental, safety and health impacts to workers and to the local public would be beneficial to all countries involved.

Research and development on DU applications

A summary of research and development (R&D) activities pursued by the countries and companies is included in Chapter V. These activities are carried out independently.

R&D projects can be divided into national and commercially-oriented projects. The former such as national repository may not involve commercial competitive factors whereas others, e.g. in the energy and industrial areas, may be affected by competition issues.

If commercially-viable large-scale uses of depleted uranium can be developed, it would be beneficial to the countries holding DU as such uses could consume appreciable quantities of depleted uranium. Although any potential large-scale uses of DU for industrial purposes would most likely be of competitive nature involving possible proprietary information, international collaboration for R&D on DU, especially during the early stage of development would be useful in fostering information exchange and minimising duplication of work. In the repository application, specific site considerations are likely to limit the extent of such collaboration.

Tracking and control of DU applications

Country-specific regulations on DU applications are potential constraints on large-scale use of depleted uranium. The need for international collaboration on further harmonisation on tracking and control of DU applications should be examined.

Public awareness

There is a lack of public acceptance for the use of DU in products which hinders its commercial viability in non-nuclear uses. Putting the radiological and chemical hazards in proper perspective would help to allay public concerns. There is a need for international collaboration on how to provide relevant information in an impartial and transparent manner.

VII. KEY ISSUES

Key issues involving management of depleted uranium are:

How long and how much should DU be stored?

DU should be stored for as long as potential economic uses exist or are perceived to exist. Currently identified usage includes re-enrichment, as energy sources, and in shielding and industrial applications. While these applications are still on-going or likely to be in the future, there is a need for continued storage of DU as a potentially valuable resource. In this period, R&D may also identify further applications. The amount retained should be determined in accordance with the expected volume of future uses.

In what form should DU be stored?

DU should be stored safely in a form most economically suitable for expected future use, subject to radiological and toxicological regulations pertaining to that specific storage site. If necessary, technologies for the deconversion of UF_6 to other forms are available.

What are the implications of final disposition of DU?

Should there be no further potential for future use there would be a need for some form of final disposition. The requirements for this will be dependent upon national policies and facilities. An international collaborative effort in sharing information and experience in this field would be beneficial to all countries involved.

REFERENCES

1. Uranium Institute (1996), *Depleted Uranium from Enrichment*, London.

2. Communication from Mr. Karel Fouché (2000), AEC, South Africa.

3. NEA/IAEA (2000), *Uranium 1999 Resources, Production and Demand*, ISBN 92-64-17198-3, OECD, Paris.

4. Lawrence Livermore National Laboratory (1997), *Depleted Uranium Hexafluoride Management Programme,* UCRL-AR-124080, Vol.1, rev.2, p.3-6.

5. Kapline Enterprises Inc. (1995), *Depleted Uranium Market Study.*

6. C.W. Forsberg and V.K. Sikka (2000), *Depleted-Uranium-Dioxide Steel Cermet for Spent Nuclear Fuel Waste Packages*, Transactions of the American Nuclear Society, Washington DC.

7. G.J. Hutchings et al. (1996), *Uranium-oxide-based catalysts for the destruction of volatile chloro-organic compounds,* Nature, Vol. 384, p. 341.

8. G.J. Hutchings; J.L. Callahan et al. (1970), Ind. Eng. Chem. Prod. Res. Dev. Vol 6, p. 134.

9. R.R. Price, M.J. Haire and A.G. Croff (2000), *Potential Uses of DU,* American Nuclear Society 2000 International and Embedded Topical Meeting, Washington DC.

10. S. Dua, M.C. Burleigh, M.J. Haire, E. Myers, and Z. Zang, and M.V. Konduru and S.H. Overbury (2001), *Putting Depleted Uranium to Use: A New Class of Uranium-Based Catalysts*, Waste Management 2001 Symposium, Tucson AZ.

11. T. Meek, M. Hu, and M.J. Haire (2001), *Semiconductive Properties of Uranium Oxides,* Waste Management 2001 Symposium, Tucson AZ.

Annex 1

GLOSSARY

Alpha radiation

Emission of an alpha particle – which consists of two protons and two neutrons, the same as the nucleus of a helium atom – from a radionuclide. It has a positive charge. More damaging than the same dose of beta or gamma radiation but can be stopped by a sheet of paper.

Decay product

The residual nucleus produced in radioactive decay; also refer to as daughter nucleus.

Depleted uranium (DU)

Uranium with a content of ^{235}U lower than the 0.711% of naturally occurring uranium.

Enrichment

Defined as isotope enrichment, i.e. increasing the content of ^{235}U. An example is enrichment of natural uranium with 0.711% ^{235}U to low enriched uranium with up to 5% ^{235}U.

Fast reactor

A reactor in which the fission chain reaction is sustained with high-speed (i.e. fast) neutrons. It is capable of converting ^{238}U into plutonium (i.e. breeding) which allows the production of a huge amount of energy from a small quantity of original uranium.

Gas centrifuge plant

Plant for isotope enrichment where uranium hexafluoride in gaseous form is introduced into centrifuges. In the field of centrifugal force, the heavier ^{238}U is concentrated at the periphery of a centrifuge while the lighter ^{235}U concentrates at the centre. Internal flow then is used to collect the enriched product at the top of the machine while the tails are collected at the bottom. This process is repeated until the desired enrichment is reached.

Gaseous diffusion plant (GDP)

Plant for isotope enrichment where uranium hexafluoride in gaseous form is forced through a barrier with small holes. The lighter ^{235}U diffuses more rapidly through these holes, thus creating enrichment. This process is repeated until the required enrichment is reached.

Highly enriched uranium (HEU)

Uranium enriched to more than 20% ^{235}U. Uranium with a ^{235}U concentration of over 90% has been produced for nuclear weapons, for reactors for ship propulsion and for research reactors. HEU can be diluted with re-enriched DU to form low enriched uranium, LEU, suitable for fabrication of nuclear fuel.

Laser enrichment

Enrichment of uranium atoms or molecules where a laser beam with specified wavelengths excites ^{235}U so that these atoms or molecules can be separated. At present laser enrichment is at the demonstration stage.

Light water reactors

A type of reactor that uses predominantly thermal (slow) neutrons to produce the chain reaction. The fast neutrons are slowed by colliding with water molecules, hence the term. The water also absorbs neutrons and so to make up for this loss, the fissile part of the uranium, ^{235}U, is enriched from natural concentration to become low enriched uranium.

Low enriched uranium (LEU)

Enriched uranium that contains between 0.711 and 20% ^{235}U. LEU is used as fuel for light water reactors, typically with ^{235}U concentrations of up to 5%.

MOX fuel

MOX fuel is a mixture of plutonium oxide and uranium oxide. The latter is usually depleted but may also be natural or slightly enriched. MOX fuel is used in light water reactors. Some plutonium isotopes such as ^{239}Pu and ^{241}Pu are fissionable. The content of ^{235}U in depleted uranium will also make some contribution to the fission in MOX.

Radiotoxicity

The ability of a radionuclide to produce injury, by virtue of its emitted radiation, when incorporated into the body.

Re-enrichment

Additional enrichment of depleted uranium tails, usually with a content of ^{235}U sufficiently high to be economical, (e.g. above 0.3%) to form LEU and depleted uranium with an even lower content of ^{235}U.

Separative work unit (SWU)

The standard measure of enrichment services, measuring the effort expended in increasing the ^{235}U content of uranium above the natural 0.711%. It typically measures the amount of enrichment services required to produce a given amount of enriched uranium from a particular feed material. Enrichment plant capacities are quoted in SWUs per annum.

Specific Activity

The activity of radioisotope per unit mass of a material, either (a) in which the radioisotope occurs, or (b) consisting of only that isotope.

Uranium hexafluoride (UF_6)

Uranium hexafluoride is a chemical compound consisting of uranium and fluorine. UF_6 is a solid at room temperature but gaseous at elevated temperatures, subliming at 55.6°C at atmospheric pressure. UF_6 reacts with moisture in the air to produce HF gas. It is stored in sealed steel containers.

Uranium oxide

A chemical compound of uranium and oxygen. There are various uranium oxides such as UO_2, U_3O_8, and UO_8. U_3O_8 is a black powder and one of the most chemically stable forms of uranium.

Sources: Uranium Institute, *Depleted Uranium from Enrichment,* London, November 1996; OECD/NEA, *Nuclear decommissioning: A proposed standardised list of items for costing purposes,* Interim Technical Document, Paris, 1999; Bodansky, D., *Nuclear Energy: Principles, Practices and Prospects,*University of Washington, AIP Press, New York.

Table A2. **Operating enrichment facilities**

Country	Location	Type	Million SWU/a	Start-up
China (1)	Shaanxi	Centrifuge	0.5	1999
France	Pierrelatte	Diffusion	10.8	1979 (2)
Germany	Gronau	Centrifuge	1.1	1985 (3)
Japan	Rokkashomura	Centrifuge	1.05	1992 (4)
Netherlands	Almelo	Centrifuge	1.5	1973 (3)
Russian Federation	Angarsk	Centrifuge	1.0	1954 (5)
	Novouralsk	Centrifuge	10.0	1949 (5)
	Seversk	Centrifuge	3.0	1950 (5)
	Zelenogorsk	Centrifuge	5.0	1964 (5)
United Kingdom	Capenhurst	Centrifuge	1.3	1976 (3)
USA	Paducah	Diffusion	11.3	1954
	Portsmouth	Diffusion	7.4	1956 (6)

(1) China has a centrifuge plant under construction in Lanzhou, 0.5 million SWU/a.
(2) Eurodif Group.
(3) Part of Urenco Group. Capacity expansion in progress.
(4) Expansion planned.
(5) Previous enrichment operations of gaseous diffusion plants.
(6) Termination of enrichment activities planned for June 2001. Plans are to maintain the plant in a cold standby condition through 2005.

The following pilot plants are currently known to be operating, on standby, or under construction: 20 000 SWU/a gas diffusion plant in Argentina, 7 000 SWU/a centrifuge plant in Brazil, 5 000 SWU/a centrifuge plant in Pakistan. In addition there are a number of R&D facilities around the world that do not generate significant quantities of tails material. The specified capacity figures are not known with a high degree of certainty.

Source: Extracted from the International Atomic Energy Agency NFCI database.

Annex 3

CHARACTERISTICS OF DEPLETED URANIUM

Depleted Uranium is mainly stored in the form of uranium hexafluoride. At ambient temperature, UF_6 is a colourless, high molecular weight (352) solid with a significant but less than atmospheric vapour pressure. It is readily transformed into a gas at atmospheric pressure by raising its temperature above 56.4°C and into a liquid by increasing the pressure above 1.5 atmospheres and the temperature above 64°C. All three phases, solid, liquid and gas, coexist at 64°C. [1]

Uranium hexafluoride is radioactive because of its uranium content and is chemically reactive because of its high fluorine content. The chemical hazards are more significant than the radiological hazards. It is a highly reactive material under conditions in which stable uranyl (UO_2^{++}) species can be formed or in which UF_6 can behave as an oxidising agent. An important UF_6 reaction is that which occurs with water to form the soluble reaction products uranyl fluoride (UO_2F_2) and hydrogen fluoride (HF), both of which are very toxic.

Radiotoxicity aspects of depleted uranium from enrichment operation[2]

The radiological characteristics of DU are a consequence of the properties of the three isotopes of natural uranium: ^{238}U, ^{235}U and ^{234}U and their daughter products. The abundance of uranium isotopes in DU varies, but it is typically 99.6999% ^{238}U, 0.3% ^{235}U and 0.001% ^{234}U.

DU is safe against criticality under all naturally occurring conditions. The initial specific activity of DU when it is newly produced is very low, around 23 kBq/g and the radiotoxicity is also low, around 0.033 mSv/Bq. Whatever the ^{235}U assay of uranium, as time passes, decay products will appear meaning that the activity and the radiotoxicity will increase to the same level as uranium ore (with the same original amount of uranium) after a time period of around 1 million years.

There are some beta and gamma emissions from isotopes of uranium and their decay products that require control in the work place. However, the external radiation hazards associated with uranium handling and storage are generally not a major concern (unless large quantities of DU are stored in small areas). Whether in the work place or in the environment, the radiological hazards from DU are primarily due to alpha particle emissions. This means that the internal radiation dose from ingestion or inhalation of uranium compounds is the limiting hazard under almost all circumstances.

One significant problem area, when working with DU, comes from finely divided airborne particles, which can result from some manufacturing operations such as machining and grinding. It is

1. IAEA, *Manual on Safe Production, Transport, Handling and Storage of Uranium Hexafluoride*, IAEA-Tecdoc-771, Vienna, November 1994.

2. Uranium Institute, *Depleted Uranium from Enrichment*, London, November 1996.

essential to provide machine ventilation, area ventilation and special filtering equipment to protect workers from radioactive dust and particles that could be inhaled or ingested.

Characteristics of depleted uranium product forms[3]

The differing characteristics of various potential DU deconversion products can have a significant impact on the acceptability of these forms for disposal. The physical and chemical characteristics of the potential DU deconversion products that are relevant to the packaging, transportation, and disposal of each DU product form are reviewed briefly in the following sections.

Physical characteristics

Table A3.1 presents physical properties of potential deconversion products. The bulk densities for the production forms of DU compounds are much lower than the theoretical densities, which are maximum values obtainable only for crystalline forms or for cast/sintered monoliths. Bulk densities are highly variable for some forms because of specific details of the deconversion process such as the mix of particle or aggregate sizes, the degree of settling that occurs or is caused to occur, and whether products are sintered to increase density. With suitable prior specification, densities near the upper end of indicated ranges appear to be achievable. Achieving a higher density is a necessary prerequisite to lowering the volume of the DU deconversion products other than metal and, thus, potentially reducing the cost of packaging, transportation, and disposal.

Table A3.1 **Physical properties of potential DUF_6 deconversion products**

Compound	Mol.wt.	Bulk density* [t/m^3]
DU metal	238	19
DUF_4	314	2.0-4.5
DUO_2	270	2.0-5.9
DU_3O_8	842	1.5-4.0

* Based on data from Duerksen et al. (2000) and Dubrin et al. (1997).

Chemical characteristics

The most important chemical characteristics of the various potential DU products are their solubility in water (high solubility enhances transport by water) and their degradation via reactions with water, which are summarised in Table A3.2.

Metal

Depleted uranium metal reacts slowly with moisture under ambient conditions to produce DU oxides and hydrogen. The oxide layer normally spalls, allowing the reaction to continue with a fresh

3. Croff et. al., *Assessment of Preferred Depleted Uranium Disposal Forms*, ORNL/TM-2000/161, ORNL, for the USDOE, Oak Ridge, USA, June 2000.

metal surface. Reaction rates in air are slow but much higher rates are observed under saturated anaerobic conditions. No detectable hydrogen is formed in the presence of oxygen except under circumstances that permit condensation of water on the metal surface and limit the transport of oxygen from the gas phase to the metal surface. There are a number of anecdotal reports of bulk uranium ignition that have been attributed to the formation of uranium hydride layers under saturated anaerobic conditions produced in storage containers [e.g. see Biwer et al. (2000)] [4]. However, an evaluation of DU metal ignition potential concludes that hydriding is not required to explain such events[5]. In addition, while the conditions that had been thought to result in hydriding could occur in both disposal and storage environments if a water layer sufficient to inhibit access of oxygen to the uranium surface is present, such conditions are not likely to be significant in arid climates and unsaturated soils.

Table A3.2 **Chemical properties of uranium and its compounds under ambient conditions**

Compound	Solubility in water	Chemical reaction
DU metal	Insoluble	• Reacts slowly with moisture to form oxides in the presence of oxygen; condensed moisture promotes H_2 generation. • Reactions may form pyrophoric surface in absence of O_2.
DUF_4	Very slightly soluble	• Reacts slowly with moisture to form DUO_2 and HF and eventually other oxides and minerals.
DUO_2	Insoluble	• Powder only can be pyrophoric in air. • Reacts very slowly with oxygenated groundwater to yield more stable oxides and minerals.
DU_3O_8	Insoluble	• Reacts very slowly with oxygenated water to yield more stable uranium minerals. • Product tends to be a fine particulate or powder.

Source: Biwer et al. 2000; DOE, *Programmatic Environmental Impact Statement for Alternative Strategies for the Long-term Management and Use of Depleted Uranium Hexafluoride*, DOE/EIS-029, Washington DC, April 1999; Duerksen et al. "Selection of optimum storage forms for depleted uranium," Appendix B in Hightower, J.R., and J.R. Trabalka, *Depleted Uranium Storage and Disposal Trade Study Summary Report*, ORNL/TM-2000/10, Jan.2000.

Tetrafluoride

Depleted uranium tetrafluoride is nonvolatile, nonhygroscopic, and only very slightly soluble in water (~40 ppm at room temperature).[6] However, evolution of fluoride ions, believed to result from

4. Biwer, B.M., et al., *Depleted uranium disposal feasibility assessment*, Appendix D in Hightower, J.R., and J.R. Trabalka, *Depleted Uranium Storage and Disposal Trade Study Summary Report,* ORNL/TM-2000/10, Jan.2000.

5. Epstein, M., et al. *On Prediction of the Ignition Potential of Uranium Metal and Hydride*, Nuclear Safety, 37 (1) Jan.-March 1996.

6. Katz, J. J., et al., *The Chemistry of the Actinide Elements,* Vol. 2, 2nd Edition, Chapman and Hall, New York, 1986.

chemisorbed HF, has been observed in stored UF_4. However, chemisorbed HF can be removed during production by heating the DUF_4 product. In addition, DUF_4 reacts very slowly with moisture at ambient temperatures to form DUO_2 and HF. If not removed, the HF will enhance the corrosion rate of the packages.

Dioxide

Finely divided DUO_2 (i.e., powder) exhibits pyrophoric behaviour. However, aggregates such as granules, pellets, and monoliths do not exhibit such behaviour. Depleted uranium dioxide is very slowly oxidised to other oxides and then to stable uranium-bearing minerals by oxygenated groundwater. Such transformations are the subject of intense study in the context of the proposed repository at Yucca Mountain, Nevada, USA.

Triuranium octaoxide

The chemical behaviour of DU_3O_8 is very similar to that of DUO_2 except that DU_3O_8 powder does not exhibit pyrophoric behaviour. However, production of DU_3O_8 tends to yield significant amounts of very fine particles, which may not be acceptable for disposal without further treatment or packaging.

Summary observations

Despite the very different chemical forms of the DU conversion products considered, all have essentially similar characteristics relevant to waste disposal. All have low to very low solubility in water, and all react *very* slowly with water to yield degradation products that are typically more stable and have a lower density (and hence greater volume) than the parent species. None of the reactions occur at a rate that would make the materials "reactive" as the term is generally interpreted in the context of managing wastes. One potential exception involves hydrogen production or pyrophoricity of hydride layers on DU metal. This concern is unlikely to be significant in an arid disposal setting. A second potential exception is the pyrophoricity of finely divided DUO_2 powder, which can be eliminated by specification of a larger aggregate as the conversion product.

Annex 4

DEPLETED URANIUM PROGRAMMES, ACTIVITIES AND PLANS AS REPORTED BY PARTICIPATING COUNTRIES AND ORGANISATIONS

CHINA

China holds stocks of depleted uranium in the form of UF_6 estimated at about 2 000 t by the end of 1999. The average depleted uranium assay is 0.25%. It is projected that the stocks will grow to 16 500 t/a by 2007.

Currently, depleted uranium is stored in UF_6 form in cylinders of the 48X and 1 m^3 types. All the cylinders are kept in reinforced steel concrete buildings. The 48X cylinders are set horizontally and in a two-tiered way. The first-tier cylinders are set on saddle formed concrete bases and the second-tier cylinders on V type wooden bases. The 1 m^3 cylinders are stored vertically in one tier on concrete ground without bases.

The buildings where the depleted uranium is stored follow specific rules on radiation protection (GB8703-88). These buildings have a normal ventilation system with air exchange rate of 3 times/hr and an accidental ventilation system with two-stage filters. Air samples are taken for radioactive measurements near the outlet of ventilation twice a month.

Depleted uranium cylinders are handled by bridge cranes or crane trucks and transported by normal 20-tonnes trucks. The 48X and 1 m^3 cylinders are manufactured according to the nuclear industry recommended standards EJ/T350 and EJ/T424, respectively. Up to the present, no defect has been identified in any of the cylinders. A 1 m^3 cylinder with over 30 years of life was cut into pieces for tests and inspection. The inside surface of the cylinder was found to be smooth and metal lustrous.

China is planning to perform feasibility studies for the conversion of depleted uranium to U_3O_8. Re-enrichment of some of the depleted uranium has been performed.

Depleted uranium is considered a potential resource for the future in China, although large-scale applications have not yet been identified. A potential long-term application is as fertile material for fast breeder reactors. In China an experimental fast breeder reactor with a capacity of 65 MW is under construction and is expected to be critical by 2005. With the expected development of nuclear power and the increase of depleted uranium stocks, China will be paying more attention to the disposition options, economics and potential applications of this material.

FRANCE

For more than three decades, France has been operating industrial gaseous diffusion plants for uranium enrichment. Two plants were built successively on the Tricastin site in the South of France: the first one, commissioned in 1965, was dedicated to defence applications; it was closed in 1996. The second plant has been producing enriched uranium since 1979 with fissile uranium content (^{235}U) lower than 5% for use in civil nuclear reactors; this plant is also called the Georges Besse plant and is owned by Eurodif, a 59.7% subsidiary of COGEMA.

About 190 000 tU of depleted uranium have been generated in France as a by-product of the uranium enrichment industry since its beginning. Most of this material originates from natural uranium. Its typical isotopic composition is:

- 0.0020% for the isotope 234.
- Within 0.20 and 0.35% for the isotope 235.
- No isotopes 232 and 236.

The depleted uranium output from the diffusion plant is in the form uranium hexafluoride. It is stored in 48 Y containers in intermediate storage yards directly located on the production site.

However, the UF_6 chemical form is not appropriate for long-term storage because it can potentially react with the water vapour in the air, producing uranium oxifluoride compounds and releasing hazardous gaseous hydrofluoric acid. The UF_6 is consequently converted into a more stable, fluorine-free chemical form, U_3O_8 which allows safer and more economical storage (smaller storage surface areas).

The conversion of the depleted UF_6 into U_3O_8 is performed at the neighbouring COGEMA W facility on the Tricastin site. The main steps of the W defluorination process are hydrolysis, followed by pyrohydrolysis using a combination of steam and hydrogen. This process has been applied on an industrial scale since 1984 and has already allowed the conversion of about 140 000 tU of UF_6 into U_3O_8. The fluorine is reduced to 70% hydrofluoric acid which is sold to the chemical industry. The U_3O_8 is compacted (density 3 to 4) and packed into DV 70 containers.

The DV 70 are three-m^3 cubic steel containers filled with about 10 t of U_3O_8. For storage, they are stacked on three levels in modular sheds, either at the Tricastin production site or at a dedicated remote intermediate storage facility (the Bessines facility in central France). Each modular shed features para-seismic qualities, has a surface area of 2 600 m^2 and can contain up to 2 200 DV 70 containers. The approval for the Bessines intermediate facility was confirmed by court decision in 1998 and storage has begun since then. The license allows the storage of up to 199 900 tonnes of depleted U_3O_8 on the site. The maximum authorised uranium mass activity is $2.11x10^4$ Bq/gU.

No final disposal option of the depleted uranium is envisioned by COGEMA for the time being because this material is considered as potentially valuable. Its ^{235}U content is still high enough to justify its re-enrichment in the case of a shortage on the uranium market. The re-enrichment operation consists of feeding the depleted uranium into the enrichment cascades. This operation has been regularly performed at the Eurodif plant.

The depleted uranium also offers several other potential interesting applications, which have been developed to a lesser or greater degree:

- Within the nuclear fuel cycle, it is used as matrix for the fuel of fast breeder reactors and for mixed plutonium and uranium oxide fuel (MOX fuel). COGEMA has a facility called TU2, on the Tricastin site, which is dedicated to the conversion of the depleted U_3O_8 into sintered UO_2. Annual quantities of about 130 tU of depleted U_3O_8 are currently used for MOX fuel. These quantities should increase to up to 250 tU/year in the short term with the expected capacity increase of the French MELOX plant.

- Outside the nuclear fuel cycle, the depleted uranium can be exploited for:
  - its physical properties (high density, high inertia), for instance in ship keels, military devices;
  - its chemical properties (colouring crystal glasses, component of some homeopathic medicines);
  - its radiological properties (neutron absorption): protection barrier for reprocessed uranium storage areas at COGEMA sites, use in the manufacturing of some heavy concrete for building surrounding walls of X-ray rooms or for storing spent nuclear fuel, etc.

Due to its low radiation level, depleted uranium requires few specific radiological protection measures; however, for prevention, there is no permanent workstation in the facilities dealing with depleted uranium. To avoid any contamination (essentially by inhalation) in the processing plants, uranium systems are all confined and a permanent control of the aerosol level in the ambient air is performed. The annual integrated dose received by the personnel of COGEMA on the Tricastin site was about 0.04 mSv for the last three years, i.e. about 100 times less than the authorised values.

REPUBLIC OF KOREA

Republic of Korea does not have any enrichment facilities and therefore does not generate depleted uranium. In addition, it is not expected that depleted uranium will be produced in the near future. However, the Republic of Korea has 16 nuclear power reactors in operation, four units under construction and eight planned by 2015. In order to obtain the required fuel for Korean light water reactors, the enrichment for the uranium has been committed to the United States or France.

Depleted uranium in the form of DUF_6 generated from enrichment is considered to be a potential energy resource for use in the long-term. If factors such as storage cost, environmental effect, technology availability and others allow, depleted uranium stocks would be reserved in the Republic of Korea to complement insufficient energy resources. In connection with this matter, the Korea Atomic Energy Research Institute (KAERI) had imported 17 cylinders of DUF_6 from the United States.

KAERI has been interested in alternative uses of depleted uranium. In the 1980s, research activities were conducted to identify potential uses including the development of containers for the shielding of radioactive material. Facilities for the conversion of DUF_6 to DUF_4 and DU metal were constructed and have operated for some time.

DUF_6

DUF_6 had been received from the United States as shown in Table A4.1. DUF_6 received in 1979 and 1983 had been used in the production of DU_4, which was further reduced to DU metal for the development of containers for the shielding of radioactive material. The remaining DUF_6, which was received in 1986 and 1988, has been stocked at the outside storage yard of KAERI. The current stock of DUF_6 is estimated to be 184.8 t.

Table A4.1 **Stocks of DUF_6 received from USA**

Year	Quantity (t)	No. of Cylinder	Type of Cylinder
1979	1.4	1	30B
1983	12.5	1	48Y
1986	73.9	6	48Y
1988	110.9	9	48Y
Total	198.7	17	

During inspection, corrosion was found on the surface, valves, plugs and skirt of some cylinders. However, the corrosion was not found to be severe and leaks had not occurred. In order to ensure protection, the cylinders were covered by special paint in 1996.

Regarding the remaining stocks of DUF_6 any concrete plans for application have not been established in the Republic of Korea. KAERI is currently seeking an adequate storage technology for the long term.

DUF_4

DUF_4 had been imported for use in research activities for the development of nuclear fuel fabrication technology, and of containers for the shielding of radioactive material as shown in Table A4.2. In addition, DUF_4 had been produced in the conversion facility from DUF_6 by KAERI. Most of the imported DUF_4 and a part of the DUF_4 produced have been used and the remaining DUF_4 is estimated to be 9.6 t stocked in the drums in KAERI.

Table A4.2 **Stocks of imported UF_4**

Year	Quantity (kg)	Origin
1980	37.9	France
1982	37.9	France
1983	1 334.4	France
Total	1 410.2	

Depleted uranium metal

In the first half of the 1980s, a technology development programme to produce U metal from DUF_4 was conducted in KAERI. Before the completion of this activity, DU metal was imported by KAERI as shown in Table A4.3. Once the reduction technology of DUF_4 to DU metal in KAERI had been developed, the amount of DU produced during 1980 to 1987 was reported to be about 1 400 kg. The depleted uranium metal has been used continuously for the R&D of radiation shielding and various nuclear fuel technologies.

Table A4.3 **Stocks of imported DU metal**

Year	Quantity (kg)	Origin
1979	880.7	UK
1986	1 995.8	USA
1988	997.0	USA
1999	1 002.6	USA
Total	4 876.1	

The remaining amount of depleted uranium metal is assumed to be about 1 000 kg. The depleted uranium scrap, which has been generated after using the depleted uranium metal as simulated material, is assumed to be about 3 800 kg.

Facilities

Deconversion facility of DUF_6 to DUF_4

The facility was constructed from 1987 to 1991 in KAERI. The capacity for DUF_4 production was about 50 t/a. DUF_4 was produced by reducing DUF_6 with hydrogen gas which was dissociated from ammonia. The generated HF gas was scrubbed with KOH solution. The facility had been used to establish the optimum operating conditions from 1992 to 1995. Unfortunately the Korean government decided to stop the projects related to the radioactive material shielding research. In addition, the DU metal facility using UF_4 was decommissioned in 1994. Accordingly this facility is in standby waiting for a decision of the responsible department in the Republic of Korea.

Reduction facility of DUF_4 to DU metal

KAERI had been particularly interested in developing the technology for making uranium alloys, which are suitable for radioactive material shielding and other applications. The technology development was initiated from 100 g/batch. The equipment capacity was increased little by little. When the largest equipment producing 100 kg/batch of DU metal entered operation in 1987, the projects supporting its operation were stopped. The facility was decommissioned in 1994. Currently, the facility is being used for the R&D of research reactor fuels.

DU metal working facility

Induction melting furnace was provided by CERCA in France in the latter half of the 1970s. Research activities using this equipment began in 1983. The various pieces of equipment related to the heat treatment and mechanical work of uranium metal had been purchased since 1984. R&D activities conducted in this facility include:

- Shielding of radioactive material container.
- Fabrication technology of research reactor fuel.
- Fundamental research of liquid metal cooling nuclear reactor fuel.

As a result a few container shields have been made preliminarily and the fuels for HANARO, which is a multi-purpose research and test reactor in the Republic of Korea, could be fabricated for the irradiation test. In addition, fuel powder samples of depleted uranium alloy produced by atomisation in this facility have been provided to BWXT in USA, CERCA in France and CNEA in Argentina.

Research activities using depleted uranium material

Radiological Shielding for containers

HANARO, which is able to produce radioactive isotopes, has been operating since 1995. The production of radioactive isotopes was initiated at that time and has increased little by little. The staff responsible for this research became interested in procuring radioactive isotope containers. Accordingly, the activities to develop the radioactive shield have been revived since 1997. As a result, some shields were fabricated successfully. In conjunction with this work, about 13 t of DU metal were scheduled to be imported in 2001.

Various fuel developments using DU metal as simulated material

As mentioned above, small quantity of DU metal has been used for the development of various fuels such as research reactor fuel and of liquid metal cooling nuclear reactor fuel. It is expected that the consumption of DU metal will be maintained.

Summary

The Republic of Korea does not have facilities generating DUF_6. However, many activities have been conducted related to the identification of future uses of depleted uranium material. Given the potential for depleted uranium to be used as an energy resource in the far future, the Republic of Korea is interested in making efforts to preserve depleted uranium stocks. Currently, the Republic of Korea does not have research activities under way related to the long-term storage of depleted uranium. If the circumstances permit, it is probable that associated research activities will be launched in the Republic of Korea.

JAPAN

A large amount of depleted uranium as tail product has arisen in Japan due to the operation of the uranium enrichment plant at Ningyo-Toge Environmental Centre. Depleted uranium stocks estimated at about 2 600 tU in the form of UF_6 are considered a potential resource for the future especially as fuel for fast breeder reactors. Until additional usage plans come into existence, its long-term storage will continue for the time being. The following are the potential usage plans under consideration by Japan:

- Shielding material for spent fuel cask.
- Application for hydrogen storage alloy.
- Application for magnetic material.
- Application for redox flow cell.

Shielding material for spent fuel cask

When designing the various storage methods for spent fuel, its nuclear properties such as gamma ray intensity, neutron ray intensity etc. should be taken into consideration. A high shielding efficiency cask is desired for the high radioactive intensity of spent fuel. Uranium, given its nuclear properties, has high shielding efficiency against gamma rays. In addition, uranium after forming uranium hydride has the possibility of being used as shielding material for neutron rays. Therefore uranium, as shielding material, is expected to be used in great quantities in some confined areas including for the storage of spent fuel.

Application for hydrogen storage alloy

Hydrogen storage alloy is a functional material for an energy storage system that uses hydrogen as a medium and that is used together with fuel cell. The working temperature of hydrogen storage and that for the dissociation of uranium are high, as well as its heat of reaction. Thus, hydrogen storage uranium alloy is likely to be used for the power storage system that applies the hydrogen generation/storage method by water electrolysis. Owing to its high heat of reaction, the power storage system that applies the hydride heat of reaction can also be anticipated.

Usage of uranium as a power storage system is advisable in some large and concentrically confined areas such as nuclear power stations or facilities related to nuclear industries.

Magnetic material

Physical properties of uranium are affected significantly by the 5f electron which is the specific electron configuration of uranium. As the 5f electron is considered to affect the uranium magnetism, the discovery of high efficiency magnetism in uranium alloys is to be anticipated.

In case of its discovery, it would be used as a substitution material for rare-earth magnets. Uranium, however, is a radioactive material, used only for limited purposes, such as permanent magnet for synchrotron, free electron laser and so on, which are usually installed in controlled areas where the radioactive materials are relatively easy to control.

Usage of the uranium alloy magnet can be recommended at facilities that have been established as controlled areas.

Development plan of uranium alloy magnet at JNC

JNC's development plans of uranium alloy magnet are as follows:

- To discover unknown intermetallic compounds by metallographic tests after melting of samples of uranium alloys.
- To carry out the magnetism measurement if unknown intermetallic compounds are found.
- To estimate the approximate structure of practical new material and continue to discover other unknown compounds.

JNC intends to carry out the above mentioned development by repetition.

Application for redox flow cell

The redox flow cell was proposed for electric power storage on a large scale. In Japan, two redox flow batteries with 200 kWe utilising vanadium have been demonstrated since 1996. The uranium is found to be of cell active material and the uranium battery is expected to have better charge and discharge performance than current vanadium batteries.[1] Therefore, the fundamental investigation of redox flow cell using uranium has been done at Tohoku University in Japan. Usage of uranium in concentrically confined areas such as nuclear power stations is envisioned.

Deconversion to U_3O_8

The basic plan is to pursue beneficial use of DU, but, on the other hand, deconversion to U_3O_8 is under consideration at JNC for long-term storage.

RUSSIAN FEDERATION

The Russian Federation holds depleted uranium stocks estimated at about 460 000 tU[2]. Most of the stocks (around 98%) are in the form of UF_6 stored in vertical containers in open-air yards. Regular control of the conditions of the containers and preventive maintenance programmes are continuously carried out to ensure adequate storage of the material especially in some cylinders with over 40 years of life.

The Russian Federation has considered different approaches for the conversion of depleted uranium from UF_6 into more stable forms. A plasma process that converts UF_6 into U_3O_8 has been developed and about 100 t of UF_6 have been processed. Another conversion process consists of low-temperature hydrolysis implemented in two steps. A pilot plant was built with a capacity of about 50 kg/h. Other processes, which have been experimentally examined, include the use of olefins for the production of UF_4 followed by hydrolysis to obtain U_3O_8 or by treatment with SiO_2 to form UO_2. The main applications identified for depleted uranium are in the production of MOX fuel and of fast reactor fuel. Additional potential applications for depleted uranium are: as feed for the production of hydrogen fluoride, as radiation protection shields, as heavy metal for flywheels, and for the production of special sorbents.

An overall programme defining the strategy, basic purposes and directions of technical activities for the storage, management and conversion of DUF_6 is being developed by the Russian Ministry of Atomic Energy, Minatom. The programme has been designed for implementation through 2010.

The programme is based on the following Russian laws and normative documents in the field of atomic energy:

- The federal law "About protection of environment".
- The federal law "About radiation safety for people".
- The federal law "About sanitary and epidemiological well-being of the population".

The concept will benefit from future improvements in the legislative and normative documents of the Russian Federation and from conformation to international recommendations.

---

1. T. Yamamura, et al., D24, 2000 Annual Meeting of the Atomic Energy Society of Japan (in Japanese).
2. Estimate based on the 1996 Uranium Institute report [1].

The implementation of the programme includes several stages:

- The first stage consists of scientific research, technical design, experimental development and economic assessment of alternative options for the storage, management and conversion of DUF_6. The necessary standard-legal documentation is drawn up.
- At the second stage, pilot testing for improvement of technology and equipment for DUF_6 conversion are performed: plasma chemical (mixed oxide and hydrogen fluoride production), reduction with unsaturated hydrocarbons (UF_4 and ozone safe khladone production), reduction with hydrogen (UF_4 and hydrogen fluoride production), etc.
- At the third stage, identified alternatives for the storage, management and use allow the production of stable and ecologically safe chemical forms of uranium such as tetrafluoride, oxides or metal.
- At the fourth stage, the profitable processing of the DUF_6 stocks is provided.
- At the fifth stage, industrial processing is implemented allowing the discontinuance of the accumulation of DUF_6.

UK/NETHERLANDS/GERMANY

The tripartite Urenco organisation was formed following an inter-governmental agreement, the Treaty of Almelo, signed in 1970, on the development and exploitation of the gas centrifuge process for the enrichment of uranium for civil uses. After construction and operation of three small pilot plants in the early 1970s, two small commercial plants were built at Capenhurst in the UK and Almelo in the Netherlands in the late 1970s. These were followed by the modular construction of further plants at these sites and, from the mid-1980s, at Gronau in Germany. Expansion of capacity is still in progress at all three sites. A centrifuge enrichment plant owned by the UK government was purchased by Urenco when it became redundant for the government's needs in the mid-1990s. Earlier a gaseous diffusion plant had also been operated by the UK government to fulfil their defence and early civil power programme needs.

The centrifuge enrichment plants operate primarily with natural uranium as feed although depleted uranium, enriched to equivalent natural uranium, and reprocessed uranium feed have all been processed in one or other of the plants. The tails generated from the early UK operations were stored outdoors in vertical 0236 containers. These are in the process of being cleaned, re-coated and moved indoors. The remainder of the tails is stored in standard 48G or 48Y containers on prepared rafts. These are periodically inspected for damage and surface rust treated.

The tails have an assay typically between 0.25% and 0.35% ^{235}U, thus retaining a significant potential for re-enrichment for use in the production of power reactor fuel. Currently it is not economic to construct new enrichment capacity to perform this re-enrichment but where spare centrifuge based capacity exists this is an economic option and some Urenco tails are currently being re-enriched.

Storage of tails in the form of UF_6 is safe for many decades and is the most economic manner to optimise the potential for early re-enrichment. At the end of 1999, 16 000 tU were stored at Urenco sites in this form. The capacity to store tails at sites is however sometimes limited by licensing and regulatory factors. In these cases preparations are being taken by obtaining licences for, and designing, above ground storage facilities for tails deconverted to U_3O_8. In Germany, there is also a further 300 tU stockpile held by ANF in Lingen. In the UK, BNFL holds 30 000 tU of depleted uranium as a valuable future resource and is investigating a plasma-based deconversion process to reduce the associated storage costs by recovering the flurorine component of the UF_6 as high value elemental fluorine gas and producing high density metal to minimise storage volumes. In such cases, storage can be continued safely for many decades maintaining the potential for supplementing/displacing the mining of natural uranium for future use in the civil nuclear fuel cycle.

In Germany depleted uranium maybe taken into consideration for a national repository, e.g. by blending spent research reactor fuel assemblies to reduce them to low enrichment level.

USA

Depleted uranium hexafluoride programme

Over the last four decades, large quantities of uranium were processed by the United States using gaseous diffusion in order to produce enriched uranium for national defence and civilian purposes. Depleted uranium hexafluoride (DUF_6) was generated as a by-product of the process and was stored at three uranium enrichment sites. A legacy of approximately 700 000 metric tons of DUF_6, containing about 476 000 tU, is currently stored at the USDOE's Paducah site in Kentucky, the Portsmouth site in Ohio, and the East Tennessee Technology Park (ETTP) in Oak Ridge, Tennessee in about 57 700 steel cylinders.

The DUF_6 is a stable but toxic, granular solid, much of which has been stored for decades. The advanced age of some of the steel cylinders in which the DUF_6 is contained, and the way in which the cylinders were originally arranged (sometimes too close together to permit inspection, and sometimes in direct contact with the ground leading to enhanced cylinder corrosion) has created a potential environmental and safety hazard. While DUF_6 does not present as significant a radiological threat as other isotopes, it is a potential chemical hazard if not properly managed.

The mission of the DUF_6 Management Programme is to safely and efficiently manage the USDOE's inventory of DUF_6 in a way that protects the health and safety of workers and the public, and protects the environment until the DUF_6 is either used or disposed. The DUF_6 Management Programme is divided into three sub-activities:

- Cylinder surveillance and maintenance.
- Conversion.
- Beneficial use of depleted uranium development.

Cylinder surveillance and maintenance

The USDOE recognises that it will take decades to convert all of the DUF_6 in the inventory. As a result, the long-term management of the existing DUF_6 storage cylinders and the continual effort to remediate and maintain the safe condition of the DUF_6 storage cylinders will remain a USDOE responsibility for many years into the future.

The day-to-day management of the DUF_6 cylinders includes actions designed to cost-effectively improve their storage conditions. This work includes:

- Collecting and managing information concerning the characteristics of the depleted uranium and cylinders including studying, monitoring, and characterising cylinder corrosion to predict corrosion breeches and identify better corrosion prevention methods.

- Performing general cylinder and cylinder yard maintenance, which includes: disposal of yard wastes (wooden saddles, construction and maintenance debris, etc.), cylinder valve changes, cylinder skirt cleaning, cylinder nameplate re-attachment/replacement, storage yard fire system maintenance, storage yard cleaning and maintaining drainage avenues.
- Performing regular inspections of cylinders to assure no degradation of their condition. Cylinders showing symptoms of accelerated corrosion are inspected annually; all other cylinders are inspected quad-annually.
- Restacking, and respacing the cylinders to improve drainage and to allow for more thorough inspections. In addition, replace wood support chocks with concrete support chocks to reduce corrosion rates and improve stability of cylinder rows.
- Repainting ends of skirted cylinders and repaint cylinders bodies as needed to arrest corrosion.
- Constructing new concrete cylinder storage yards and reconditioning existing yards from gravel to concrete. This will improve storage conditions, drainage, and permit restacking of cylinders to reduce cylinder corrosion rates, damage from handling, and allowing more thorough inspections.
- Dispositioning current inventory (empty and heel) DUF_6 storage cylinders.
- Obtaining US Department of Transportation certified overpacks to allow for the off-site transportation of the 48-inch DUF_6 storage cylinders.

Conversion

Since 1990, the DUF_6 Management Programme has largely focused on the ongoing surveillance and maintenance of the cylinders containing DUF_6. The question, however, of the long-term management and eventual disposition of DUF_6 remains and is the subject of considerable interest within the Congress, the USDOE, and with concerned citizens and stakeholders.

The US Congress stated its intentions with regard to DUF_6 in the Public Law (P. L.) 105-204, signed by the President in July 1998. This law directed the Secretary of Energy to prepare and submit to Congress a plan for the construction and operation of plants to treat and recycle the DUF_6 consistent with the National Environmental Policy Act (NEPA).

In the fourth quarter of calendar year 2000, the USDOE issued a request for proposals to design, construct, and operate conversion facilities on the Department's property at Paducah, Kentucky and Portsmouth, Ohio. These facilities will convert the USDOE's inventory of DUF_6 to triuranium octoxide (U_3O_8), uranium dioxide (UO_2), uranium tetrafluoride (UF_4), uranium metal, or some other stable chemical form acceptable for transportation, beneficial use/reuse, and/or disposal. Any of the proposed conversion forms will be required to have an assured, environmentally acceptable path for final disposition. Cylinder surveillance and maintenance will become the responsibility of the conversion contractor as soon as practicable after contract award. The selected contractor will also be responsible for transportation of the ETTP cylinders to Portsmouth for conversion.

Beneficial use of the depleted uranium

The USDOE has completed the development of a *draft Depleted Uranium Uses Roadmap*, a plan that will be used in managing the disposition of its DUF_6 inventory. The Roadmap characterises and analyses alternative paths for the eventual disposition of surplus depleted uranium, makes recommendations on which paths are preferred, identifies the barriers that exist for those paths, and proposes research, development, and other activities in order to eliminate the barriers. The *draft Depleted Uranium Uses Roadmap* was made available for public review and comment in the fourth quarter of fiscal year 2000. Comments have been received and are being incorporated. When completed, a final version of the Roadmap will be approved and posted on the Internet.

In addition, the USDOE has initiated the Depleted Uranium Uses Research and Development Programme to explore the potential beneficial uses of the depleted uranium (DU), fluorine, and empty carbon steel DUF_6 storage cylinders to achieve cost savings to the government.

The programme's key goals are to:

- Support a broad spectrum of investments to reduce the barriers to paths that have relatively low technical risk and use large quantities of DU in regulated areas related to nuclear material storage and/or disposal;
- Make targeted investments to reduce barriers for a number of paths where there is potential to use substantial amounts of UF_6 conversion products but where the uses are more speculative or simply require a small investment before the path could be followed;
- Make appropriate investments to ensure that there are no barriers to following an optimal path for long-term storage or disposal of the DU that is not beneficially used or DU-bearing devices at the end of their useful lives; and
- Invest in basic and mission-directed research that is related to beneficial use of UF_6 conversion products.

Building on the initial results of the Roadmap a number of tasks were initiated related to uses of DU as a shielding material, catalyst, and as a semi-conductor material.

The USDOE is committed to the beneficial use of the depleted uranium and other materials resulting from the conversion of the DUF_6, i.e., fluorine and contaminate carbon steel storage cylinders, for the purpose of achieving benefits, including cost savings to the government compared with simply disposing of the materials. However, the government intends to assure that direct disposal of these materials is possible if cost-effective and realistic beneficial uses cannot be found.

Annex 5

MEMBERS OF THE EXPERT GROUP ON POTENTIAL USES AND MANAGEMENT OF DEPLETED URANIUM

China

Mr. Yuansong Liu China National Nuclear Corporation Department of Nuclear Fuels P.O.Box 2102-10 BEIJING 100 822	Tel. Fax: E-mail	+86 10-6853 1390 +86 10-6853 1355 liuys@cnnc.com.cn

France

Ms. Brigitte Le Motais COGEMA Boîte Postale 4 F-78141 VÉLIZY Cedex	Tel. Fax E-mail	+33 1 39 26 3371 +33 1 39 26 2721 blemotais@cogema.fr
Ms. Marie-Thérèse Ménager CEA - Direction de la Stratégie et de l'Évaluation 33, rue de la Fédération F-75752 PARIS	Tel. Fax E-mail	+33 1 40 56 1719/2927 +33 1 40 56 1514 marie-therese.menager@cea.fr
Ms. Anne Guinot COGEMA, Boîte Postale 44 F-26701 PIERRELATTE Cedex	Tel. Fax E-mail	+33 4 75 50 4643 +33 4 75 50 7224 aguinot@cogema.fr

Germany

Mr. Wolfgang Thomas GRS GmbH Postfach 1328 D-85739 GARCHING	Tel. Fax E-mail	+49 89 32 004 490 +49 89 32 004 491 tho@grs.de

Japan

Mr. Ippei Amamoto Japan Nuclear Cycle Development Institute (JNC) Kamisaibara-son 1550 Tomata-gun OKAYAMA 708-0698	Tel. Fax E-mail	+81 868 44 2211 +81 868 44 2566 amamoto@ningyo.jnc.go.jp

Mr. Eiji Suzuki
Japan Nuclear Cycle Development Institute (JNC)
Kamisaibara-son 1550, Tomata-gun
OKAYAMA 708-0698
Tel. +81 868 44 2211
Fax +81 868 44 2566
E-mail esuzuki@ningyo.jnc.go.jp

Mr. Osamu Takanobu
Facility Management Division
Ninyo-Toge Environmental
1550, Kamisaibara-mura, Tomada-gun
OKAYAMA-ken 708-0698
Tel. +81 868 44 2211 ext. 6131
Fax +81 868 44 2833
E-mail tentaka@ningyo.jnc.go.jp

Republic of Korea

Mr. Chang-Kyu Kim
Korea Atomic Energy Research Institute
P.O.Box 105, Yusong
TAEJON 305-600
Tel. +82 42 868 2309
Fax +82 42 864 1089
E-mail ckkim@nanum.kaeri.re.kr

Mexico

Mr. Ruben Ramirez Guerrero
Comision Nacional de Seguridad Nuclear
y Salvaguardias de Mexico
Dr. Jose Maria Barragan, No 779, piso 3
033020 MEXICO, D.F.
Tel. +52 50 95 3232
Fax +52 50 95 3293
E-mail dg1@servidor.unam.mx

Russian Federation

Mr. Victor Seredenko
Russian Research Institute of Chemical Technology (RRICT)
33 Kashirskoe Avenue
MOSCOW 115230
Tel. +7 095 324 4050
Fax +7 095 324 5441
E-mail chem.conv@g23.relcom.ru

United Kingdom

Mr. Robert J. Gunn
Department of Trade & Industry, Room 108
1 Victoria Street
LONDON SW1H 0ET
Tel. +44 20 7215 2785
Fax +44 20 7215 2843
E-mail robert.gunn@dti.gs.gov.uk

URENCO (United Kingdom, Netherlands, Germany)

Dr. Stephen. J. Baker
Urenco Limited
18 Oxford Road
MARLOW, Bucks, SL7 2NL
Tel. +44 1628 402 254
Fax +44 1628 402 202
E-mail sjb@urenco.com

United States

Mr. Larry Dewey *(Chairman)*
Office of Nuclear Energy, Science and Technology, NE-20
US Department of Energy
1000 Independece Avenue
Washington, DC 20585
Tel. +1 202 586 4290
Fax +1 202 586 2027
E-mail larry.dewey@hq.doe.gov

Mr. Robert Price
Office of Nuclear Energy, Science and Technology
US Department of Energy
19901 Germantown Road
GERMANTOWN, MD 20874

Tel. +1 301 903 9527
Fax +1 301 903 4905
E-mail robert.r.price@hq.doe.gov

Consultant

Ms. Sashi Davies
ZLS Consulting Limited
2 Cherry Walk, Sarrat Lane
RICKMANSWORTH, Herts WD3 6AF
United Kingdom

Tel. +44 1923 77 1598
Fax +44 1923 77 3539/1406
E-mail sashimdavies@aol.com

International Atomic Energy Agency

Mr. Jor-Shan Choi, *(IAEA Secretary)*
IAEA Nuclear Fuel Cycle Materials - A2626
P.O.Box 100
A-1400 VIENNA

Tel. +43 1 2600 22771
Fax +43 1 26007/2600 29598
E-mail j.s.choi@iaea.org

OECD Nuclear Energy Agency

Dr. Ivan Vera, *(NEA Secretary)*
OECD Nuclear Energy Agency
Nuclear Development Division
Le Seine St-Germain
12, Boulevard des Iles
F-92130 Issy-les-Moulineaux

Tel. +33 1 45 24 1063
Fax +33 1 45 24 1110
E-mail ivan.vera@oecd.org

ALSO AVAILABLE

NEA Publications of General Interest

2000 Annual Report (2001) *Free: paper or Web.*

NEA News
ISSN 1605-9581 Yearly subscription: FF 240 US$ 45 DM 75 £ 26 ¥ 4 800

NEA Brochure *Free: paper or Web.*

Geologic Disposal of Radioactive Waste in Perspective (2000)
ISBN 92-64-18425-2 Price: FF 130 US$ 20 DM 39 £ 12 ¥ 2 050

Radioactive Waste Management Programmes in OECD/NEA Member Countries (1998)
ISBN 92-64-16033-7 Price: FF 195 US$ 33 DM 58 £ 20 ¥ 4 150

Nuclear Development

OECD Energy Data 2001 (2001)
Bilingual
ISBN 92-64-08707-9 Price: € 20 US$ 19 £ 12 ¥ 1 900

Methodologies for Assessing the Economic Consequences of Nuclear Reactor Accidents (2000)
ISBN 92-64-17658-6 Price: FF 200 US$ 31 DM 60 £ 19 ¥ 3 250

Business as Usual and Nuclear Power (2000)
ISBN 92-64-17175-4 Price: FF 160 US$ 25 DM 48 £ 16 ¥ 2 850

Reduction of Capital Costs of Nuclear Power Plants (2000)
ISBN 92-64-17144-4 Price: FF 240 US$ 38 DM 72 £ 24 ¥ 4 400

Nuclear Education and Training: Cause for Concern? (2000)
ISBN 92-64-18521-6 Price: FF 210 US$ 31 DM 63 £ 19 ¥ 3 300

Nuclear Education and Training: Cause for Concern? (2000)
A Summary Report (booklet) *Free: paper or Web.*

Uranium 1999: Resources, Production and Demand (2000)
ISBN 92-64-17198-3 Price: FF 510 US$ 77 DM 152 £ 48 ¥ 8 100

Nuclear Energy in a Sustainable Development Perspective
ISBN 92-64-18278-0 *Free: paper or Web.*

Actinide and Fission Product Partitioning and Transmutation (2001)
Sixth Information Exchange Meeting, Madrid, Spain, 11-13 December 2000 *Free: paper or Web.*

Order form on reverse side.

ORDER FORM

OECD Nuclear Energy Agency, 12 boulevard des Iles, F-92130 Issy-les-Moulineaux, France
Tel. 33 (0)1 45 24 10 15, Fax 33 (0)1 45 24 11 10, E-mail: nea@nea.fr, **Internet: http://www.nea.fr**

Qty	Title	ISBN	Price	Amount
		Total		

❑ Payment enclosed (cheque or money order payable to OECD Publications).

Charge my credit card ❑ VISA ❑ Mastercard ❑ Eurocard ❑ American Express

(N.B.: You will be charged in French francs).

Card No.	Expiration date	Signature
Name		
Address	Country	
Telephone	Fax	
E-mail		

OECD PUBLICATIONS, 2, rue André-Pascal, 75775 PARIS CEDEX 16
PRINTED IN FRANCE
(66 2001 13 1 P) ISBN 92-64-19525-4 – No. 52057 2001